Whole Energy System Dynamics

In order to address the twenty-first-century challenges of decarbonisation, energy security and cost-effectiveness it is essential to understand whole energy systems and the interconnection and interaction between different components. An integrated language is therefore needed to assist energy policymakers and to help industrial stakeholders assess future energy systems and infrastructure and make realistic technical and economic decisions.

Whole Energy System Dynamics provides an interdisciplinary approach to whole energy systems; providing insights and understanding of it in the context of challenges, opportunities and solutions at different levels and time steps. It discusses approaches across disciplinary boundaries as well as existing issues within three main themes: theory, modelling and policy, and their interlinkage with geopolitics, markets and practice. Spataru argues that there is an urgent need for a whole energy system integration. This is necessary for effective analysis, design and control of the interactions and interdependencies involved in the technical, economic, regulatory and social dimensions of the energy system.

This book is essential reading for students interested in the area of energy systems, policy and modelling. It is also a valuable read for policymakers, professionals, researchers, academics, engineers and industrial stakeholders.

Catalina Spataru is a Lecturer in Energy Systems and Networks at UCL Energy Institute in London, UK, Course Director of MRes in Energy Demand Studies at UCL and Group Leader of Energy Space Time Group. Her expertise is in whole energy system dynamics, with great interest for regional interconnections (EU, West Africa, MENA countries, Latin America) and market integration, future energy development in emerging economies, reliability and security of future energy systems and with particular interest in the interaction of water–energy–land nexus to assess water scarcity problems in risky countries like Brazil, Jordan, Egypt.

She regularly delivers presentations in academic and professional circles, public engagement events and for the media, recently visited Princeton University and MIT (USA), spoke at Cheltenham Science Festival and was interviewed by the *Sunday Telegraph*. Currently she teaches Smart Energy Systems for MSc EDE students at UCL, a multidisciplinary module that she proposed and developed in 2013. This multidisciplinary module provides students with an understanding of the methods, concepts and practice of whole energy systems, offering a combination of theory, modelling interactive exercises and case studies. Also she teaches Communication Skills to MRes EDS students and Metrics, Modeling and Visualisation of the Resource Nexus module for the ISR MSc sustainable resources programme. She supervises MSc, MRes and PhD students. She is the regional representative of the IEEE Women in Power (Region 8 – Europe).

'*Whole Energy System Dynamics* provides a genuine means of understanding the energy system by identifying three main pillars: the theory of interacting systems, the practice of modelling the whole energy system and the policy that steers its future evolution. The book rises to the challenge of integrating the numerous parts of today's energy systems while connecting the different disciplines and thus allowing for various perspectives on the complex systems we are all part of.'
Andreas Koch, Group Manager, Energy Planning and Geosimulation, EIFER – European Institute for Energy Research, Germany

'This book comes at a time when there is an urgent need to have a complete rethink of our future energy options. It cuts through the ingrained perceptions hindering our move from fossil fuel to sustainable forms of energy. This book, which integrates theory, practical perspectives, modelling and policy aspects and links traditional engineering ways with the very latest developments, will be a beacon in our quest for a secure and sustainable energy future.'
Trevor Letcher, Emeritus Professor of Chemistry at the University of KwaZulu-Natal, Durban, South Africa and a Fellow of the Royal Society of Chemistry

'This book is well written, hugely interesting and an excellent format for discussing new problems, particularly for people working on energy systems and policy. It provides profound analysis and understanding of the energy systems integration from theory, practice/modelling and policy perspectives as a whole. I enjoyed reading it.'
Pierluigi Siano, Professor of Electrical Energy Engineering, Electrical Power Systems, University of Salerno, Italy

'This book is very timely as whole system energy modelling is one of the hottest research topics at the moment. Research funding agencies in many countries have recognised the need to research and model the energy system as a whole and provided ample funding to support it. This is a relatively new area of research and there are few books available on the subject. This book therefore fills an important gap in the market and is relevant not only for researchers but also for policy makers.'
Janusz Bialek, FIEEE, Director of Skoltech Center for Energy Systems, Russia

Whole Energy System Dynamics

Theory, Modelling and Policy

Catalina Spataru

First published 2017
by Routledge
2 Park Square, Milton Park, Abingdon, Oxon OX14 4RN

and by Routledge
711 Third Avenue, New York, NY 10017

Routledge is an imprint of the Taylor & Francis Group, an informa business

British Library Cataloguing-in-Publication Data
A catalogue record for this book is available from the British Library

Library of Congress Cataloging-in-Publication Data
Names: Spataru, Catalina, author.
Title: Whole energy system dynamics : theory, modelling, and policy /
Catalina Spataru.
Description: Abingdon, Oxon ; New York, NY : Routledge is an imprint of the
Taylor & Francis Group, an Informa Business, [2017] | Includes
bibliographical references.
Identifiers: LCCN 2016025043| ISBN 9781138799899 (hb) | ISBN 9781138799905
(pb) | ISBN 9781315755809 (ebk)
Subjects: LCSH: Power resources. | Energy industries. | Energy policy.
Classification: LCC TJ163.2 .S694 2017 | DDC 333.7901/1--dc23
LC record available at https://lccn.loc.gov/2016025043

ISBN: 978-1-138-79989-9 (hbk)
ISBN: 978-1-138-79990-5 (pbk)
ISBN: 978-1-315-75580-9 (ebk)

Typeset in Sabon
by GreenGate Publishing Services, Tonbridge, Kent

Contents

Illustrations

Figures

Tables

Preface

Introduction

Over the years I have worked on various multidisciplinary projects and experienced at first hand the rift between perspectives and viewpoints among scholars belonging to different disciplines in different parts of the world. I have many times faced unending disputes between scientists, engineers and economists.

My main conclusion is that the three pillars of theory, practice/modelling and policy have to be clearly understood and integrated. They need to be recognised together as an interconnected whole instead of independent parts. The theory will help in gaining insight as to how well it fits the evidence through practice and to meet policy requirements. This necessitates the need of modelling to gain understanding of different component parts of the energy systems.

Modelling whole energy systems, however, is a great challenge mainly due to the inherent uncertainty present in the systems. The fact is that energy systems are massive, complex and in a state of constant flux and development. Energy demand is driven not just by the conscious decisions, but also by ingrained behavioural patterns of consumers. Add in the element of technology, and the process of energy systems modelling becomes increasingly complex.

Technologies have evolved greatly in the past few decades, and will continue to evolve over the subsequent periods of decades and centuries. The evolution is guided by innovation and infrastructure requirements, as well as geopolitics and economics. And this change is happening with great inertia. Trying to understand whole energy systems with traditional tools without taking into account the latest advancement in technologies is like getting insight into the Big Bang using simple tools of Newtonian mechanics.

Traditional engineering tools are useful to gain insights into specific parts of the problems apropos energy systems. However, they are not adequate for understanding the whole energy systems and the related challenges. The traditional tools don't possess the depth required for detailed analysis of the systems. They need to be complemented by the latest tools from multiple domains.

The provocative challenge is to make students understand the use of mathematical models to describe the reality and provide quantitative and qualitative answers to questions raised by practitioners and policymakers. Students usually are afraid of learning modelling or they don't find it important – which also intrigues me. My challenge over the years has been to show students, researchers and practitioners how well the theories that guide our thinking, acknowledged or not, fit the facts and how models using the theory describe the reality.

The book that you are reading right now was written with the main aim of getting better insights and understanding of whole energy systems in the context of challenges, opportunities and solutions. Only through detailed understanding of each of the three pillars – theory, practice and policies – can we be able to comprehend the challenges and obstacles that are present in the way of sustainable and safe utilisation of our precious energy resources.

Time to share and integrate

I have divided this book into three sections that I believe are also the three pillars that will provide deep understanding of the current and future developments of the energy system:

a) *Theory*: summarises the background and guidelines for the construction of models;
b) *Practice*: brief historical overview of the implementation of energy systems and comparison of actual performance with model guidance; and
c) *Policy*: outcomes of the performance in light of the implementation of current energy policies, connections with regulations and geopolitical impact.

The central argument of this book is that there is an urgent need for a whole energy system integration. This is necessary for effective analysis, design and control of the interactions and interdependencies along with the technical, economic, regulatory and social dimensions of the energy system. This book is a small endeavour to understand the current dynamics of the energy system, which will hopefully contribute to future development and improvements of the process.

Theory

The interactions and interdependencies among the physical scale of the energy system, as well as between the energy system and other systems (e.g. data and information networks), increase significantly. Modern energy systems represent an evolution from small, local, basic service systems such as the steam engines that fuelled the Industrial Revolution to highly integrated, continental systems that deliver energy services including natural gas and electrical transmission/distribution systems to our homes and businesses.

Practice

Urbanisation, modernisation and economic and population growth are placing ever-increasing pressure on energy systems at a global level and forcing us to squeeze more out of less. Building large-scale infrastructure such as high-voltage transmission is increasingly difficult and particularly costly in many parts of the world. Moreover, the seemingly incessant advancements in technology are driving energy demands at a local level. These two factors have increased the need to optimally integrate and control limited natural resources across multiple scales – from the local level to the global energy system.

The increase in complexity of the energy system dynamics today has resulted in greater challenges in maintaining stable, reliable and cost-effective operation. In this context, we need to consider how the energy system can be made robust so that it can effectively cater to present needs and potential demands in the future.

At the moment there are various uncertainties; for example, how the new technologies will integrate and impact from a social and economic perspective. Uncertainty relating to fuel costs can have significant impacts in the transportation and power generation sectors. Extreme events and unforeseen changes in the energy system landscape such as rapid development of US shale gas and re-entry of the oil-rich country Iran into the international arena have resulted in beneficial effects on the economy due to cheap fuel oil, but at the same time have been disruptive to the ecology as a result of greater usage of 'dirty' fuel. This necessitates the need to develop energy system integration as a major research area with the key goal of proposing solutions that will allow development of a robust energy system that can adapt to a range of unknowns.

A whole energy system and integration of its components is theoretically simple but fundamentally complex to implement due to the presence of a great degree of interdependencies that are difficult to appreciate; the complexity arises from the technical difficulties across space and scale, intertwined with social, political and economic factors. Without considering the interconnection between different factors, it is highly probable that a solution will be provided that will fail to provide an optimal solution. In addition, global optimisation may lead to a brittle solution with consequences in terms of security and reliability. However, a set of optimal subsystems may improve global results and resilience. To do this, a good understanding between subsystems and the interactions across them is needed. It is in our collective best interest to guide research towards a solution that can deliver a range of feasible futures. This can be done through collaborative research between industry and research centres/universities by integrating scalable modelling and simulation tools in space and time with large sets of data.

Policy

Thomas Edison is recognised for leading the world into the age of electricity. He used to say that 'the greatest compliment I heard in my entire life as an inventor was: it will never work'. How many times have we heard phrases like that when working on implementing a seemingly impossible task?

'It will not work...', 'It will take time to do...'

Now I realise that these words are the manifestations of distrust and concern about the scale of the project that can have a significant impact on multiple sectors once implemented. It was the fear of disruptions or disturbances that placed doubt on the public supply of electricity. And it is the same doubt and distrust that blocks the implementation of effective policies that will help in utilisation of clean, affordable sources of energy that does minimum harm to the environment.

To sum up

The three parts of the book form the very foundation that can contribute to further development of energy systems in the future. Each of the three parts of the book is subdivided into a number of chapters.

Part I presents

A historical overview of energy systems theory with an overview of different aspects that steered many of the steps that are adopted in the modern energy model.

Part II provides

A brief introduction to modelling history and practice while providing a number of examples at different scales and levels with a list of open sources. This part of the book discusses how to apply energy system theory to practice.

Part III explains

The foundations of the energy system policy with a brief overview on global system policies, together with the functions, challenges and interfaces. It also describes and analyses the bases of coupling markets, presents different aspects of the implementation process, and offers some conclusions and remarks at the end within the geopolitical concept.

The essential dependencies between the parts indicate the importance of the subject, highlighting some specific methods, applications and challenges. The concepts included in this book cover all scales from residential to continental, across multiple energy (and non-energy) domains, i.e. the electricity and gas nexus, electricity and water nexus, gas and transport and others, from a technical, market and regulatory perspective.

Learning from the past we can understand how to change the future. Human use of energy has grown enormously, based on burning fossil fuels that has caused a significant change in the composition of the atmosphere, which can have very serious consequences if we don't do something now to change it. Energy, both as heat and work, has played a central part in the development of human societies throughout the world.

The Gaia Hypothesis can be summed up as: 'The earth is a self-regulating environment. All the living organisms and the inorganic material of the planet are part of a dynamic system that regulates the conditions to support life.'[1] This can tell us that we should consider the whole system, not just parts of it, because everything is interconnected, and paying attention to only one element of the system can have unforeseen consequences.

Furthermore, nature is not built up from isolated blocks, but from a complex web of interdependent relationships. The integration at different dimensions (micro-, meso- and macro-levels) from the financial to technical and social is becoming a necessity to find the most economic transition path.

Multi-scale modelling of energy resources, technologies, policies and control systems is crucial to understanding the dynamics of energy and society. This requires adaptation of a whole system approach with the goal of genuine sustainability and security.

Energy security has a number of dimensions: changes in the global distribution of demand and supply, increase in the import of fossil resources, political focus on the national control of supply and production, affordability of energy import for low income countries and micro-level access to an affordable and reliable supply. In this context, there is an increasing need for global interconnectedness to deal with energy issues. There is a growing focus on enhancing both global and regional dialogue to address this interconnectedness.

Future research directions

Models of energy systems or components have played a key role in formulating energy policy. However, as we move from developing strategies to implementing, there is a need for a new generation of energy system models particularly at a system level for a number of reasons.

Connecting disciplines

Whole system models enable the full representation of demand side and supply side changes to be tested and integrated across sectors, so that the impact of the interaction of policies in different areas on each other can be fully tested. In the domain of energy system modelling there is a need for a better integration between physically based models of the energy system and econometrically based models.

Perhaps more important than the epistemological hierarchy is the fact that many interesting problems are mostly viewed as primarily physical issues with economic consequences that can be worked through as an addendum to the physical analysis. Their key features can be expressed more quickly, powerfully and elegantly in the language of physics and engineering than in the language of economics.

Rapid reductions in emissions imply rapid changes in both infrastructure and behaviour. Periods of very rapid change tend to render price information less useful or unreliable. Economic analysis can obscure the sometimes very simple results of the physics-based analysis of such problems.

Conversely, many such problems are very effectively viewed from the viewpoint of mass flows through and into infrastructure, and are therefore ideally suited to physics-based analysis. For both these reasons, there is an urgent need to develop physics-based models.

Representing multiple timescales and viewpoints

As we move to implementation in system dynamics there is a real need to understand the transient impacts better. Many system models look at annual or monthly energy use but not at short-term phenomena over days or minutes.

As we move away from fossil energy sources, the need for physics-based analysis to lead in the area of the transient analysis of systems arises from the fact that energy storage costs for electricity are orders of magnitude greater than for oil, gas and coal.

Allowing for multiple viewpoints

People rather than buildings use energy to maintain health, comfort and productivity. From this viewpoint a notion of service rather than infrastructure-driven system models emerges. People can only occupy one space at any one time, and there are massive potential savings to be had from just conditioning occupied spaces.

Most system models do not allow these factors to be fully investigated and yet some of the easiest energy wins may come from moving to a person-based rather than infrastructure-based approach to analysing the energy system. Radical shifts in modelling perspective can probably be achieved more easily and transparently with physics-based models.

With regard to the economic system models, the conventional economic analysis of prices includes the concepts of price elasticity of demand and supply, both of which are nonlinear functions (power laws) of changes in price. This formulation, coupled with time dependence and asymmetry in elasticity for positive and negative changes in price, are sufficient conditions for the emergence of chaotic behaviour in energy prices. This could help to:

- understand the nature and extent of such instabilities;
- investigate the possible impact of global oil and gas production peaking;
- relate instability in simulations to instability in the real world; and
- begin to explore the possibility of designing energy systems to minimise objective functions based on both price and price stability.

Note

1 J. E. Lovelock (1979). *Gaia: A New Look at Life on Earth*. Oxford: Oxford University Press.

Acknowledgements

I would like to express my greatest gratitude to a number of people without whom I would not have been able to write this book. My biggest thanks goes to Anya and Andrei for their love, support and understanding; and to my parents for bringing me into the world.

I would like to express my gratitude to my mentors and colleagues from whom I have learned so much, for their encouragement during my career at UCL, especially to Prof. Mark Barrett, Prof. Raimund Bleischwitz, Prof. Tadj Oreszczyn and Prof. Bob Lowe. I would also like to thank my many collaborators over the years for fruitful discussions and debates, my co-authors of articles and reports. I apologise for not including all their names but I am afraid there are too many.

Because of the interdisciplinary nature of the topic I was supported by my researchers (in alphabetical order) who contributed research support on literature and data as follows: Tola Adeoya (West Africa region and review of models), Priscila Carvalho (Brazil and Chile) and Eleni Zafeiratou (review of models), and I wish to thank all MSc/MRes students at UCL who attended my lectures as part of my module Smart Energy Systems over the years, or whom I supervised over the years, for their engagement in discussions and analysis and their passion for the subject. Without them there would have been nothing to write about. Their questions and discussions made me determined to write this book. I want to thank everyone who provided encouragement and enthusiasm for writing this book on such a complex topic and for what I was trying to do by providing valuable advice. I would also like to thank my publishers.

This book has been a huge challenge due to the interdisciplinary scope and the combination of research and practical experience. Many thanks to all, and if any errors remain then I apologise in advance.

Abbreviations

AC	alternating current
AIM	Asia-Pacific Integrated Model
ASG	Asian Super Grid
BESOM	Brookhaven Energy System Optimization Model
BoP	Balance of Plant
BRICS	Brazil, Russia, India, China, South Africa
CAT	Centre for Alternative Technology
CCCA	Camden Climate Change Alliance
CEE	Central East Europe
CERC	Clean Energy Research Center
COP	coefficient of performance
CSC	current source converter
CWE	Central Western Europe
DC	direct current
DMG	Distributed Multi-Generation
DNO	Distribution Network Operator
DSIM	Digital Station Intelligence Manager
DSM	demand side management
DSM	distributed storage method
DTT	Demographic Transition Theory
ECN	Energy Research Centre of the Netherlands
EPS	Energy Policy Simulator
EVs	electrical vehicles
FEA	Federal Environmental Agency
FIP	Feed-In-Premium
FIT	Feed-In-Tariff
FSEC	Florida Solar Energy Center
GDP	Gross Domestic Product
GMM	Global MARKAL model
HDI	Human Development Index
HPs	heat pumps
HVDC	high voltage direct current
IEA	International Energy Agency
IEO	International Energy Outlook
IGBT	Insulated Gate Bipolar Transistor
ISEP	Islington Sustainable Energy Partnership

LEAP	long range energy alternative planning
LPG	liquid petroleum gas
MC	market coupling
MENA	Middle East and North Africa
MESAP	Modular Energy System Analysis and Planning Environment
MNC	multinational corporation
MoU	Memorandum of Understanding
MS	Member State
MTDC	Multi-Terminal Direct Current
NEMO	Nominated Electricity Market Operator
NEMS	National Energy Modelling System
NEOP	Non-End of Pipe
NEW	north-western Europe
NREAP	National Renewable Energy Action Plans
NWPP	Northwest Power Pool
OECD	Organisation for Economic Co-operation and Development
OPEC	Organization of Petroleum Exporting Countries
PCA	Principal Component Analysis
PCR	Price Coupling of Regions
PEM	polymer electrolyte membrane
PGCIL	POWERGRID Corporation of India
PHS	pumped hydro storage
PII	policy impact indicator
POLES	Prospective Outlook on Long-term Energy Systems
PSU	public service utility
PtG	power-to-gas
PWM	Pulse Width Modulation
PX	Power Exchange
RAI	remuneration adequacy indicator
RES	renewable energy sources
RESGEN	Regional Energy Scenario Generator
SAGE	System for the Analysis of Global Energy Markets
SGM	Second Generation Model
SPF	seasonal performance factor
SSG	SuperSmart Grid
TC	transmission capacity
TGC	Tradable Green Certificate
TSO	Transmission System Operator
UCPTE	Union for the Coordination of Production and Transmission of Electricity
UCTE	Union for the Coordination of the Transmission of Electricity
UKMD	UK Marketing Database
UN	United Nations
UNSC	UN Statistical Commission
VoFEN	Vision of the Future Energy Networks
VSC	voltage source converter
WAPDA	Water and Power Development Authority
WEO	World Energy Outlook
WPD	Western Power Distribution

Part I

Theory and history

Prologue

The global energy system is in a state of flux and is gradually moving towards a more reliable and clean source of energy. The framework for the change has been provided by significant contributions and integrations of the past relating to energy systems.

An energy system represents the integration of different physical interactions with a focus on practical and viable energy solutions for closing the gap between the demand and supply of energy to society. Any contributions made today in energy research and operations will serve as a blueprint for the development of energy systems in the future.

Energy systems have undergone significant changes since the 1800s. Many fiction writers such as seventeenth-century French writer Jules Verne predicted the evolution of energy systems. The historical timeline reveals that breakthrough inventions that contributed to the development of energy systems such as accumulators and photovoltaic cells had been foreseen by Verne.

Advancements in energy system technologies have contributed to the development of other fields. Advanced energy solutions have spurred development in the communication and transport sector through scientific advances that have led to the invention of superior and reliable technologies.

The energy system today represents a great advancement as compared to the past. It comprises multiple components that together make up the energy mix that includes electrical energy, nuclear energy, wind energy, thermal energy and solar energy. All these sources have different levels of efficiency, costs and risks that contribute to the demands on the energy systems. Moreover, these components have different interdependencies that broaden the scope and usages.

In the first part of the book we will take a look at how the different components of energy systems can be utilised effectively in different sectors, and the extent of the anticipated change that we can expect in the mix in the near future. As mentioned previously, current advances in energy systems have been made through the contributions (theoretical and practical) made in the past, which have led to the discovery of sophisticated solutions in the domain of energy systems. This is the intention of the book: we will look at the components and models of the current energy system and analyse current inefficiencies in the system and the deficits in terms of the global energy supply.

Despite advances in energy systems, it is evident that limited success has been achieved in terms of extra-efficient and sustainable energy systems as projected by Verne and other early fictional scientists. It is, therefore, important to conduct effective research and analysis of the deficiencies in the current energy systems, and design more sustainable and reliable energy models that minimise costs as well as reduce environmental impacts.

1 Global energy systems

A historical perspective

All living beings that inhabit this world depend on different sources of energy to survive. The three most common energy sources that animals and plants use for their sustenance are sun, water and food. Without energy, no life form can exist in this world.

Human beings have been using energy sources since time immemorial. Cavemen used fire to cook food, warm themselves and also frighten their enemies. Gradually more sources of fuel for energy were found such as wood, animal dung, charcoal and others that directly contributed to the development of human civilisation. The Sumerians, the Assyrians, the Egyptians, the Greeks and the Romans all used advanced technologies such as windmills and water-wheels, grinding grain and pumping water, that utilised wind, water and sun as energy sources.

With time, various other energy systems were developed that resulted in increased human productivity and economic growth. In this chapter we will take a look at the historical timeline to trace key factors that have led to the development of modern energy systems.

1.1 The three pillars of human development and progress

a) Economic growth across space and time

A look at the historical timeline will reveal that the world economy experienced significant growth after the 1500s. This can be partly explained due to the advancement in international trade and transport technologies. Economic growth was also high during the period 1870 to 1913 mainly due to the Industrial Revolution as well as further advancement in transport technologies.

From 1914 until 1945, the global economy experienced a gradual growth fuelled by three important factors: rapid rise in population, fewer barriers to international trade, and breakthrough technological advancements.

Economic growth brings prosperity in a country, which leads to improved quality of life for the general populace. We can measure economic growth through different metrics. The most popular benchmark of economic growth is gross domestic product (GDP) per capita, which represents an estimation of all goods and services produced by a country in one year divided by the country's population. The measure of the change of GDP from one year to the next is called economic growth.

Various books written by a number of contemporary writers have examined economic changes over time and in different regions through reconstructions of GDP per capita. One notable book that I would like to briefly mention here is *Monitoring the World Economy 1820–1992*, written by Angus Maddison, who was a well-known British economist.

Maddison was born in 1926 in Newcastle upon Tyne, United Kingdom. He spent most of his life working as an economics professor at the University of Groningen, Netherlands, and died in 2010 in Neuilly-sur-Seine, France. After his death, his colleagues continued his work in the 'Maddison Project'.[1] His book, *Monitoring the World Economy*,[2] is a fascinating and stimulating read that covers the entire world economy over the past two thousand years. He brought together data from around 56 countries representing 93 per cent of world output. Based on his study, the evolution of world GDP is shown in Figure 1.1.

You can see in Figure 1.1 that a significant economic expansion took place from 1950 to 1973. Average per capita world GDP since 1820 is depicted in Figure 1.2.

Maddison's study found that there was a huge difference among regions and countries along with gross inequities in the distribution of income between individuals.[3]

Another book of Maddison titled *The World Economy: A Millennial Perspective* explains and explores the factors that contributed to the success of rich countries and the obstacles that hindered other countries that lagged behind.[4] The book quantified long-term changes in world income and population. According to the study, three interactive processes have driven advances in population and income over the past millennium. These are:

a) Conquest or settlement of relatively empty areas which had fertile land, new biological resources, or a potential to accommodate transfer of population, crops and livestock;
b) International trade and capital movements;
c) Technological and institutional innovation.

Maddison says that during the past millennium the world population increased 22 times while at the same time per capital income levels increased by 13 times and world GDP an amazing 400 times. This is in stark contrast to the GDP growth experienced in the preceding millennium when there was no significant increase in population and economic growth. The above listed three processes contributed to growth in the human population and economy.

b) Human global population growth over time and space

World population growth has never been constant either in time or space: the growth was either positive or negative. Increase in rate of world population can be expressed by a simple equation as follows:

Change in Population Density = (Births + Immigration) – (Deaths + Emigration)

Let us look back in time to better understand and comprehend the human population growth and possible limits. Worldwide population was somewhere around 2 to 20 million in about 8000 BC; while by AD 1, the worldwide population had increased to 200 to 300 million people. During the 1500s, the human population had reached 400 to 500 million. It took about a millennium and a half for it to increase twofold in size.

Since 1500, the world population has seen an exceptionally quick development. The year 1820 represented a significant point in history as the world human population reached around a billion.[5,6,7]

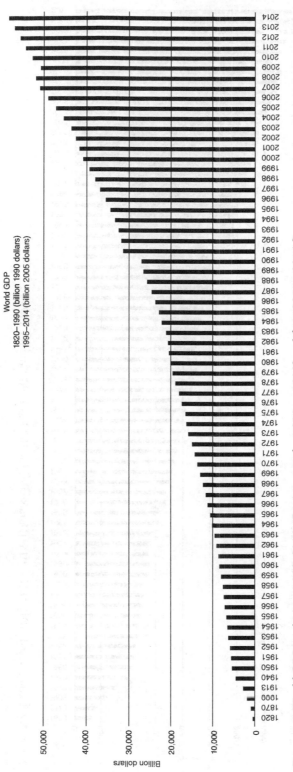

Figure 1.1 World GDP evolution, 1820–2014. (Data source for 1820 to 1990[8] and for 1995 to 2014.[9])

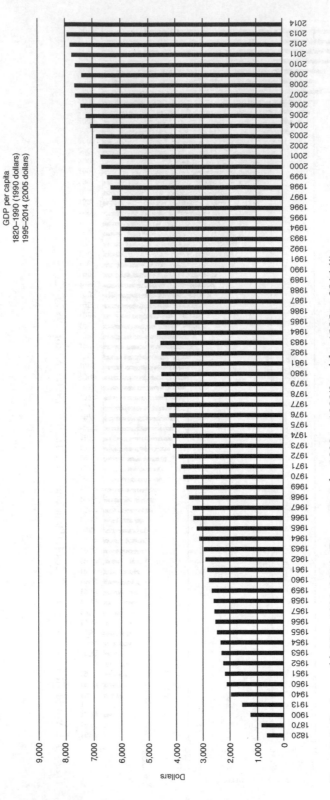

Figure 1.2 Capita world GDP, 1820–2014. (Data source for 1820 to 1990[10] and for 1995 to 2014.[11])

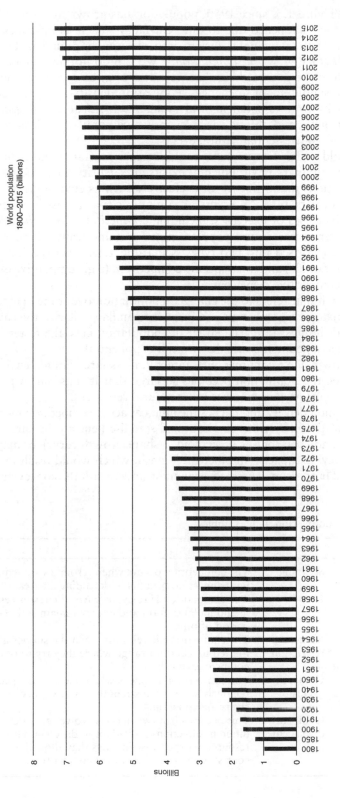

Figure 1.3 World population. Data source: 12,13

As you can see in Figure 1.3, since 1950, population has grown at a very high rate. This fast development happened due to different factors such as upgrades in well-being, diet and so forth. A valuable insight into this can be found in Livi-Bacci's book *A Concise History of World Population*.[14] We can do nothing about the past; however, what's to come is something that we can change as a consequence of our behaviour, and we should adapt to changes. Population density in a particular location is influenced by four essential natural events: *birth*, *death*, *immigration* and *emigration*. Other biotic and abiotic variables influence population density through one or more of these four essential natural events.

For instance, cold winter temperatures could increase mortality and reduce population density. The auxiliary environmental events are: density-independent factors, i.e. events or conditions that influence all exclusively such as environmental change; and density-dependent factors, i.e. events or conditions that adjust in severity as a population's size increases or decreases, including *competition for limited resources* – the individual offer decreases as a population grows numerically. Competition for resources can be categorised into two types (see Table 1.1): intra-specific competition (e.g. contest, scramble) and inter-specific competition (e.g. competitive exclusion, range restriction, competitive displacement).

Population growth that refers to the change in population over time in a particular place can be calculated using a compound interest formula similar to the calculation of a bank account's interest rate. The larger the population gets, the faster it grows. Population occurs at a geometric rate that we call exponential.

Exponential growth can't continue indefinitely in a resource-limited environment. A time comes when the population becomes so large that it runs out of free space, exceeds its food supply, and/or exhausts other natural resources.

Once the population density approaches the limit capacity, competition increases. In this case, the population may either level out and stabilise beneath the carrying limit, which is a concept known as logistic or sigmoid (S-shape growth curve), or may rapidly overshoot the conveying capacity and thereafter crash, which would result in repeated cycles of boom and bust, or may oscillate around or underneath the conveying limit.

Table 1.1 Types of competition with examples

Competition types		Details
Intra-specific	Scramble	The competition occurs when resources are temporary with almost no benefits in defending an area.
	Contest	The competition is relevant when the resources including land, food and energy remain stable over a period of time.
Inter-specific	Range restriction	The competition is relevant when the species are limited to a certain range where they try to out-compete each other.
	Competitive exclusion	The competition happens when one type of species is relatively superior in strength or intelligence, and drives rivals to extinction.
	Competitive displacement	This occurs when two species evolve separately in different directions and adapt to different natural resources or specialise in ways that allow them to coexist with no or little direct competition.

However, I believe that these models proposed by experts are rather over-simplified. In reality, natural populations react to a variety of environmental conditions that are rarely constant over time. Be that as it may, if exponential utilisation of resources is not represented in population planning, disaster can happen.

Let's do some simple calculations of population growth for the year 2020. The population value for mid-year 2000 is 6.102 billion.[15] With the following simple formula we can approximately calculate the projected world population for any year.

$$N = N_i e^{Kt}$$

where

N is the future value for population
N_i is the initial value, in this case the population for the year 2000
K is the rate of growth
t is the number of years over which growth is to be measured.

So, if we choose the year 2020 and consider the annual world population growth rate 1.36 per cent, then the population will be calculated as follows:

$$N = (6.102 \times 10^9) \times e^{0.0136 \times 20} = 8.009 \times 10^9 = 8.009 \text{ billion}$$

In the late eighteenth century, an English philosopher named Thomas Malthus wrote in his work *An Essay of the Principle of Population* about what he saw as the dire future of humanity.[16] According to Malthus the unquenchable urge to reproduce would ultimately lead us to overpopulate the planet, eat up all its resources, and die in a mass famine.

Two important questions arise at this point: What is the optimum capacity of the Earth to offer sustenance to the people? And how long will it take the population to push the capacity of the Earth to its limit?

Demographic Transition Theory (DTT) contrasts greatly with Malthus' theory in answering the population quandary. Developed by the American demographer Warren Thompson in 1929, DTT consists of the following main points:

- Fertility is not taken for granted, it is socially constructed.
- Population does not grow uncontrollably. It rather stabilises at a slow growth rate as birth and death rates fall as a country transitions from a pre-industrial to industrialised economic system.
- Absolute population growth is still large due to a large base.

The Limits to Growth,[17] which was a Neo-Malthusian study report on the predicament of mankind, generated unprecedented controversy when it was published in 1972 due to its predictions of the eventual collapse of the world's economy. The study was the first to use dynamic modelling to make predictions about future economic prospects.

On behalf of the Club of Rome, Donnella Meadows, Dennis Meadows, Jørgen Randers and their team worked on systems analysis at Jay W. Forrester's Institute at MIT. They created a computing model for simulating alternative scenarios taking into account various global developments and their relationship. They took into account different possible amounts of available resources, levels of agricultural productivity, environmental protection and birth control. Their study reveals that the world population and economy will continue to grow until the year 2030. World population and wealth would remain at a constant level if drastic measures to protect the environment are taken to change the system behaviour. In this study no account of political measures was considered.

The book has been revised by Ugo Bardi, and a second version of the book appeared, titled *The Limits to Growth Revisited*.[18] Bardi examined whether the methods and approaches of *The Limits to Growth* can contribute to an understanding of what happened to the global economy in the Great Recession, demonstrating growth scenario-building using system dynamics models.

Another scientist, the population biologist Joel Cohen from Columbia University, said in *Life's Little Mysteries* that nobody knows when or at what level peak population will occur.[19] Numerous different variables restrict the Earth's carrying capacity: the nitrogen cycle, available amounts of phosphorus, and environmental carbon concentrations.

Every two years the United Nations Department of Economic and Social Affairs' Population Division prepares the official United Nations estimates and projections of world, regional and national population size and growth, and geographic indicators. As noted in the recent report *World Population Prospects – The 2015 Revision*,[20] the current population growth (1.18 per cent per year) is slower than ten years ago (1.24 per cent). According to the report, the world population is projected to reach 8.5 billion in 2030, 9.7 billion in 2050 and 11.2 billion by 2100. As with any type of projection, there is a degree of uncertainty. The results are based on the medium projection variant, which assumes a decline of fertility for countries where large families are still prevalent as well as a slight increase in countries with fewer than two children per woman on average. The uncertainty surrounding the median trajectories is accounted for with statistical methods.

In any case, sooner or later we can make a U-turn. Empirical data from 230 countries since 1950 shows as per the UN that the considerably dominant part has a falling fertility rate at the substitution level, which is 2.1 children per woman. This is the rate at which children replace their guardians and compensate for the individuals who die young. If this is the case, then before the century is over the human population could settle somewhere around 9 to 10 billion.

We should perceive to what extent it will take for a given population at a specific growth rate to achieve a density of one person for each unit of the Earth's surface area. To do such an estimation the following expression is utilised:

$$t = (1/K) \; ln(N/N_i)$$

where

t is time in years
K is the growth rate
N is the population density
N_i is the starting population.

Considering the same scenario as above, the year 2000 with a world population growth rate of 1.36 per cent, and considering the total land mass of 1.48×10^{14} m^2 representing about 30 per cent of its total surface area,[21] we obtain the following:

$$t = (1/0.0148)ln((1.48 \times 10^{14}) / (6.102 \times 10^9)) = 682 \ yr$$

So, 2000 + 682 = 2682, which means in the year 2682 a population density of 1 *person / m^2* would be reached. However, there are already places where population densities approach 1 *person / m^2*.

In 2002, the eminent Harvard University socio-biologist Edward O. Wilson published the book *The Future of Life* where he proposed that the requirements of the biosphere are settled.[22] As Malthus argued over 200 years earlier, there are constraints on the amount of food that the Earth can produce. Wilson believes that if everybody agreed to become a veggie lover, leaving little or nothing for animals, the current 1.4 billion hectares of arable area (3.5 billion acres of land) would support about 10 billion people. He concluded the maximum carrying capacity of the Earth, taking into account the food resources, will probably miss the mark concerning 10 billion.

From my knowledge these studies do not take into account other ways of growing plants without soil such as hydroponics. Hydroponics is the term used to describe ways of growing plants by the use of an inert medium – sand, gravel, peat, pumice, sawdust, vermiculite, crushed rocks or bricks, shards of cinder blocks, styrofoam – to which is added a nutrient solution containing all the essential elements needed by the plant for its normal growth and development.

Based on history, the process of hydroponics is an ancient technique. A number of scholars are of the view that the famous Hanging Gardens of Babylon were irrigated using this technique. They suggest that the garden, considered one of the ancient Wonders of the World, was watered using a complex hydroponic system. The water pumped into the garden was rich in nutrients and oxygen, due to which the plants of the garden were green and lively. Other examples of hydroponic culture are found in China (floating gardens), as described by Marco Polo in his famous journal *Floating Gardens of the Chinese*.[23]

Much later, in 1940, William Frederick Gericke, a professor at the University of Berkeley, California, published *The Complete Guide to Soilless Gardening*[24] about innovative technology, which was written during the time when The Third Reich had occupied Paris and Hitler, Mussolini and Hirohito signed documents to form the Axis alliance.

Gericke wanted to end the war by multiplying food production. Instead of using nutrients to improve soil quality, Gericke suggested using hydroponic systems with different combinations of nutrients and growth conditions. Its prototype system's tomatoes, flowers and lettuce led to record yields that exceeded virtually the whole agricultural output in that time on conventional land. The US Army used the technology mentioned by Gericke during the Second World War to feed troops stationed on islands in the middle of the Pacific Ocean.

After the Second World War, the commercial use of hydroponics expanded worldwide in countries such as England, Germany, Italy, Spain, France, Sweden, the USSR and Israel. The high interest in this ancient yet relatively novel system was because no soil was needed. The system could be used in those very small areas that have a large population and the products had a much longer shelf life.

Another more important point is that growing plants in a non-soil medium allows them to thrive in a limited space. Moreover, crops will mature more quickly and provide greater yields, while the fertiliser and water are conserved as they are reused.

If we do have enough water and energy, solutions such as this can help maintain worldwide population growth successfully.

c) Energy growth (history)

Until the Industrial Revolution, natural forces, from human slaves and domesticated animals, to wind and water, chemical energy stored in wood and different biomass, were used as sources of energy.

Sun and wood were the main energy sources for humans. About 5,000 years ago humans started utilising wind for transportation purposes. Around 2,500 years later, windmills and water wheels began to be utilised to granulate grain, and later to pump water and run sawmills. Early Egyptians utilised oil collected from the top of lakes for lighting, while the ancient Chinese utilised natural gas to heat sea water for salt, by piping the gas from shallow wells. Geothermal energy began to be utilised at about the same time for heating houses.

The first natural gas was dug in 1821 and primary oil in 1859 (see Figure 1.4). Later, in 1882 in New York, Thomas Edison built the first power plant. Pearl Street Power Station sent electricity to 85 buildings in the state. The first gasoline-powered car was built in 1892.

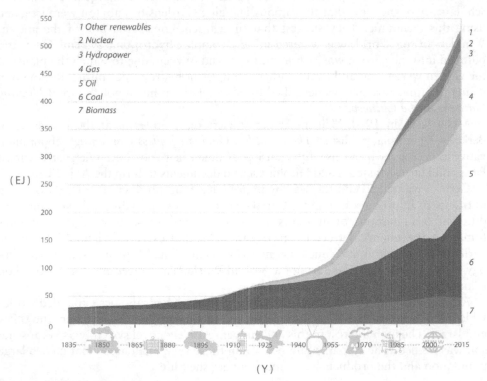

Figure 1.4 World primary energy use.

In the eighteenth century, steam engines burnt coal to convert chemical into mechanical energy. Initially, steam engines were notoriously inefficient, losing more than 99 per cent of their energy. By 1900, the same engines were 30 times more efficient than a century earlier, and they could be utilised on ships and railroad trains, permitting the transport of coal at an enormous scale.

In 1850, a Scottish chemist named James Young discovered how to refine crude oil. Later, in 1859, the American Edwin Drake demonstrated that oil could be drilled through deep rock. With this momentous discovery, the oil age had begun.

After the 1880s internal combustion engines were developed mainly in Germany with significant upgrades in effectiveness on a small scale and having more power than steam engines on a larger scale. From 1900, biomass, coal and oil were the principal sources of energy.

The electrification of the globe started around 1890. Lenin broadly characterised communism as electrification of Soviet power. Rural electrification in the US was a noteworthy accomplishment during Franklin D. Roosevelt's administration.

The utilisation of fossil fuels led to an increase in pollution and increased inequality in wealth and influence in various parts of the world. Harnessing fossil fuels played a central part in enlarging international wealth and increasing the income inequality that is so prominent in present-day history.

At the moment there is no risk of depletion of fossil fuels on a worldwide scale. Predictions about the exhaustion of fossil fuel sources have proved to be false since the 1860s. Current predictions, which are revised continuously, infer quite a few decades before oil or gas supplies run out. The issue today is whether we should keep on relying on these assets or turn to alternative fuel sources that do less harm to the environment.

1.2 People in time and space

We spend the vast majority of our time and energy in 'boxes' of various shapes, either stationary boxes (buildings) or versatile boxes (cars, trains, planes and so on). The basic unit is the cell that could be anything such as a vehicle cabin or a room. Rooms can be grouped into dwellings including houses, office buildings or schools. Dwellings and premises in turn are clustered into blocks of flats, office blocks and so forth. Vehicle cabins are clustered into trains and double decker buses.

Activities of people change with space and time. This is valuable information that defines the requirements of buildings such as hot water, space heating, lighting and cooking. People generally complete tasks in groups in spaces including living rooms, classrooms and workplaces. The use of time varies by gender, age, wealth, occupation, etc. That being said, most of us spend our time inside enclosed spaces, and most of our needs are fulfilled by systems that are powered by energy.

Throughout the years, utilisation of energy has increased significantly. In the event that service frameworks are intended to meet the administration needs inside of a specific space when occupied, then energy consumption could be diminished.

In a dwelling a large fraction of the rooms are empty, even when the house is occupied. If rooms are designed as thermally isolated cells within a dwelling, energy consumption could be reduced. Service systems could be refined to give a more exact delivery in space and time. This would require changes to the dynamic components of systems, and to passive elements. For example, concerning passive elements, it is unrealistic to deliver heat precisely in space, if internal partitions are inadequately insulated, or in time, if the effective thermal mass is large.

Possible changes to the dynamic components of systems could be heating just the regions utilised (e.g. a sofa if occupants are sitting in front of the TV); lighting the areas in use through utilisation of systems that distinguish where individuals are looking by having directional lumières that just light that part of the room; and using taps that just supply water when hands are present.

Moving boxes (transport) are high consumers of fossil energy, creating high quantities of CO_2 and being significant contributors to air pollution. Noteworthy changes are important in order to reduce carbon emissions and air pollution. This goal can be achieved by using clean energy by connecting photovoltaic panels to battery storage, and axial small wind turbines to moving boxes.

1.3 Spatial scale, location and time scale

The deployment of integrated energy solutions varies with location, spatial scale and technical design choices. It is also strongly dependent on the existing infrastructure and its resilience to supply and demand variability, as well as on current practices and regulations.

a) Spatial scale

The main challenge in the development of integrated models is capturing the dynamics and interaction between multiple spatial scales at the small and large scale levels. An issue with current models is that they are limited to a single spatial scale, e.g. a neighbourhood, city or district.

In reality, energy consumption depends on different variables at different spatial scales that need to be taken into account when optimising the urban energy infrastructure. In current practice, energy system representation typically focuses on a main scale and depicts macro phenomena in the form of exogenous assumptions.

Scale, for instance, is typically a pre-requirement that is mostly inherently assumed for the estimation of energy usage with time. Depending on assessment scale, the hourly load curve can differ a lot but at the same time show a robust smoothing effect due to enhancement of the system limit. When modelling systemic interactions, this approach falls short of delivering a dynamic system model.

Considering both spatial and temporal resolution, the number and type of selected components to be connected have great technical, financial and environmental impacts due to the sizing of infrastructure, choices, technologies and the necessity of operating the whole system. A further advantage of larger scales results also from the distributional effects of different components, with systemic influences from independent or stochastic influences, suitable for reducing the insecurity inherent in the energy assessment.

b) Location

A technology that has been implemented successfully in one region may not enjoy the same success in another region. Due to the uneven distribution of natural resource deposits across large areas, the pattern of the use of resources is quite different from one region to another.

Climate change, as well as increasing population, puts growing pressures on important natural resources. This necessitates a change in our perception about sustainable development and the management of resources.

The resource nexus, which includes not only energy but also water, sits at the intersection of many critical economic, political and social issues. The complex and interwoven connection issues require analysis and policy approaches that explicitly recognise these connections. Depending on the specifics of a given energy system in a specific location, a portfolio of solutions to minimise the risks to the system and the costs can include the development of centralised versus distributed generation (or both), as well as a strong extension of the network infrastructure versus more off-grid alternatives.

c) *Time scale*

Energy demand varies temporarily according to social activity patterns (home, work and travel), demand management (insulation levels, space heating requirements, lighting, cooling) and weather (ambient temperature, wind, solar radiation).

Energy demand thus varies throughout the year and on a sub-hourly basis. The entailed variability of demand impacts the technical and economic requirements on the energy supply side, which today relies on a larger share of renewable energy (also variable!) as well as storage, back-up capacity and energy trade as buffers.

The whole service supply system from people through to end-use technologies must be considered in order to comply with security and safety requirements, as well as the usability and adaptability of the entire system. In terms of modelling and simulation, this means favouring the use of systems-of-systems to multiply the decentralised decision processes, adopting a large number of heterogeneous entities and improving the communication and interactions between them.

1.4 Historical background since the concept of the 'whole' energy system started (1800–)

The energy system has been in constant transformation with breakthroughs made in the past two centuries. These advances were imagined in the fictitious narrations of various writers well before the actual discovery was made. For instance, Michel and Jules Verne predicted that people in the twenty-ninth century would live in a fairy wonderland where devices run on electricity would rule the lives of individuals.[25]

The energy sector has experienced significant changes in the past 250 years. The changes have been based on the theoretical projections of various researchers who have made significant contributions to the transition.[26] Although some of the advances made in the field may seem to be non-significant, they have nonetheless refined the process of delivering energy solutions to people. By looking back at the advances made in the past few centuries, it is possible to make some projections about future developments in the sector.

In the nineteenth century, Michel and Jules Verne indicated that towns in the future would have different population sizes, highly developed streets, and houses 1,000 feet high featuring constant temperature throughout the seasons.[27] The authors also anticipated that the lines of aerial locomotion would fill the skies in all directions. When writing their fictitious narrative, they would never have thought that their projections would one day come true. They lived in an era where most houses were made of mud and the roads were filled with animal-drawn carriages. A time would come when all of these would become obsolete, and the world would be dominated by modern electrical inventions and advanced transport and communication tools.[28]

Considering the predictions made by Michel and Jules Verne, it is evident that most of them have come to be true. They provided the most accurate representation of the current energy system which represents a marked improvement over past technologies. Comparing the underlying projections of modern-day achievements in the energy, transport and communication sectors, it is evident that there are similarities between centuries-old projections and current energy technologies. The development of energy started with a pleiad of innovators who made admirable discoveries that provided the backdrop for advances in energy systems.

The Vernes had stated that the modern accumulators would be instruments that stored a lot of energy, including energy contained in waterfalls, streams and the wind, and absorbed and even transformed solar energy, to meet energy demands. Their concept of modern accumulators was self-sustainable in that they could harness energy from natural sources. The class of accumulators could transform power into different forms, such as heat, mechanical force, light and electricity, to be used to perform different tasks.[29] Most of these predictions have come true today.

That being said, the predictions made by the Vernes are far from being attained, as the current advances in energy storage are still primitive. Modern accumulators can only store transformed electric energy and release this energy when required. They are designed in a way to accumulate low power energy and release it whenever required, but they cannot generate or transform the energy.[30] The current accumulators are still inferior as they can only store energy in specific forms, including mechanical or thermal, and deliver the energy in the same form. The developments are evident all the way from the steam accumulators, hydraulic accumulators, batteries, capacitors, compulsators and pumped-storage hydroelectric plants.

For example, modern batteries today can only store energy and cannot transform it into other forms of energy. Research and inventions are still ongoing to produce the next phase of accumulators, but the progress has been made on a limited scale.

Current achievements in energy are still way behind those proposed by Michel and Jules Verne when considering the development made in this field from the Neolithic to the present time. From the Neolithic period until the late eighteenth century, human beings relied mostly on fossil fuels as their source of energy. Today only slight improvements have been made to burning technology, including for wood and organic materials.

Of all the different forms of energy, coal has been used for the longest duration by humans.[31] This energy source is thought to have been in use for performing different tasks for the past four millennia, and it is still being used as a source of energy for heating and cooking purposes. The earliest known use of coal was recorded among the Aztecs during the early second century. Moreover, early evidence also suggests the use of coal in China in 2000 BC.

During medieval times in Europe the use of coal was evident and was referred to as 'the black stone'. However, since coal produced a lot of smoke and soot, the use of coal as a source of energy was replaced during the early thirteenth century by wood.

The use of natural gas as a source of energy was recorded in 200 BC in China. The Chinese used natural gas for making salt from brine. Natural gas was used as a fuel for evaporators that used bamboo pipes to deliver gas to the heating chambers.

Elsewhere in the world people were thinking of new and cleaner sources of energy. In Europe, for instance, people were using water energy to power mills for grinding grain, tanning leather and spinning cloth. In addition, Europeans rank among the earliest population to have used water energy for industrial uses that resulted in increased overall industrial productivity.

The use of human physical labour decreased significantly after 200 BC, and this is attributed to the innovative discoveries in *alternative sources of energy*.[32]

Development of alternative sources of energy paved the way for the early development of industries and promoted industrialisation in Europe as well as other major cities around the world. Due to the introduction of innovative alternative energy sources, overall economic productivity increased, and led to reduced dependence on animal and human labour.[33]

Advancements in the use of water energy technologies provided a platform for novel and effective uses of water energy. Apart from the use of water as a source of energy, other energy sources were also becoming prevalent in the world. During the first century AD, the Chinese became among the first people to refine oil for use as an energy source.[34] The discovery of oil was a remarkable breakthrough that would later take the world by storm, and greatly transform major sectors of the global economy.

Similar to coal, oil has been used as an energy source for thousands of years, making it one of the earliest known forms of energy. The earliest exploration of oil is recorded in north-western China when the inhabitants discovered oil seeping through the rocks. Moreover, a book by an anonymous historian titled the *Hand Book Geography Annals* mentions the use of flammable oil by residents along Weshio River in the present-day city of Ji'an in China.[35]

Wind is another significant source of energy that has been used since ancient times. The earliest known use of wind to power a machine is a wind wheel that was invented by a Greek engineer to power an organ. However, a major breakthrough in wind technology was made in the ninth century AD with the invention of horizontal windmills that were used for pumping water and milling grain in the Kingdom of Persia.[36]

Windmills were mainly comprised of vertical or horizontal carousel-type structures that used wind energy to rotate grinding stones for crushing grain and lifting water to gardens for irrigation purposes.

With time, the development and use of wind energy for different functions spread to other areas including India and China. Due to this innovation, more peasants could now tap the energy for different on-farm applications. Use of wind power reached the epitome of its development with the complex windmill designs developed by Dutch engineers in the early sixteenth century.

The new designs made by the Dutch inventors were developed to harness large volumes of wind energy to perform complex tasks such as grinding larger volumes of grain, pumping water for irrigation purposes, and even sawing wood. The new mills were more developed than the Persian ones and protected offshore lands from inundation caused by strong wind currents from the sea.[37]

The discovery of the new use of wind energy opened the ways for novel uses of wind energy and new developments in modern wind-powered machines. Significant development in terms of mechanisation and energy use came with the invention of the first steam engine in the year 1792.

Steam engines were initially used for pumping water out of coal mines. It is important to note that coal remained the sole source of energy during the pre- and post-industrialisation period. However, since the mines were running deeper and deeper, there was a problem with water flooding the mines.

Despite advancements made in wind-powered pumps, the competition for the most popular source of energy was won by steam-powered pumps, which were developed by a European innovator and engineer named Thomas Newcomen in the year 1712.

A steam-powered engine could pump up to 10 gallons of water out of the mines. The first steam engines were primitive and inefficient, but nevertheless they were able to assist coal miners to dig deeper than before. It is estimated that by the year 1700 Europe was using about 2.7 million metric tons of coal, which was more than 20 times the value of energy that could be produced by existing woodlands.

Thomas Newcomen's machine was the starting point that led to the development of machinery that did not rely on direct energy from animals, water or wind. The invention of the steam-powered engine brought a new era to Europe's industrial expansion.[38] The scientific invention of the steam-powered engine paved the way for the invention of new and innovative machines, and marked the point of economic and industrial revolution in Europe. Steam energy played a significant role in the Industrial Revolution in Europe that in particular helped to exploit the abundance of coal resources for use in industrial processes.

Production and use of coal expanded to other regions such as the US, with the first mines being opened in Richmond, Virginia. The colonisation of America played a major role in the establishment of mines, as noted by a number of historians.[39] The coal mines in Virginia had a significant geopolitical relevance especially during the US War of Independence and other revolutionary wars fought in America.

During the 1800s a lot of significant discoveries and advances were made in energy systems development including the increased use of natural gas and oil. The US government made significant efforts in exploiting the use of natural gas, which was initially used by the Chinese. It was in 1821 that the first natural gas well was drilled in the US in the state of New York by a geological explorer named William Hart.

Hart started digging the 27-feet well after noticing bubbles of gas and wanting to tap the gas and use it as a source of energy. Thus, Hart is referred to as the father of natural gas exploration and initiated the modern development of the use of gas for domestic purposes.

Electric power as a source of energy represents another major breakthrough in the development of energy systems. A significant breakthrough in the use of electricity came in the year 1800 with the discovery of electrolysis by British scientists William Nicholson and Anthony Carlisle.[40]

Discovery of the process of electrolysis was also the first step towards the use of hydrogen energy where hydrogen gas that is produced by the reaction between water and electricity is used to generate electricity. Use of hydrogen gas as an energy source represents the next stage of the clean energy initiative being pursued today.

During the nineteenth century, natural gas was the major source of domestic lighting in most parts of the world.[41] However, due to lack of the required piping infrastructure, gas was mainly used near the source of production. Other than that, it was during this time that major discoveries were made regarding electric current. In addition, it was during the 1830s that ethanol was blended for use as lamp fuel in most homes in the United States.

It is important to note that long before the discovery of petroleum the major source of fuel for heating was mainly 'camphene', which is composed of ethyl alcohol and turpentine with drops of camphor oil added. Camphene had a peculiar smell and was the preferred burning fluid in many homes at the time. The use of ethanol blends patented in 1834 replaced the reliance on expensive whale oil and marked the onset of the commercial production of alcohol for lighting.[42]

The development of fuel cells in 1838 by Robert Grove was another significant contribution in energy system development that paved the way for the use of direct current as an energy source as scientists used platinum and zinc electrodes to generate electricity. It was also during the 1800s that Nicholson and Carlisle successfully decomposed water into its constituent elements: hydrogen and oxygen.[43]

Modern energy systems aim to replace the current reliance on fossil fuels for internal combustion engines. Combustion engines that can electrolyse water and utilise hydrogen for combustion may become common in the future as they emit oxygen which results in minimum harm to the ecosystem.

Meanwhile, advancements in wind technology continue to receive significant support from governments worldwide. Wind today remains an important energy source for countries around the world. This important renewable energy source was perfected by Daniel Halladay in the year 1850 through the establishment of Halladay Windmill Company that was the first commercial company to provide windmill construction services.[44]

Initially, the windmills were mainly used for pumping water for the Union Pacific steam locomotives subsequent to the completion of the railroad construction. It was around 1850 that windmills were commercialised in America and across the world; in America alone there were estimated to be around six million windmills used for pumping water into homesteads. The search for better energy sources was still on, and the development of crude oil use became actualised in the year 1859 when Abraham Gesner developed a method for distilling oil from crude petroleum.[45]

Modern oil refinery concepts are based on Gesner's discovery. The discovery of the oil distillation process opened a new chapter for petroleum exploration and mining, with Edwin Drake drilling the first oil well, and this marked the commercial shift towards the use of petroleum as a domestic energy source. The exploration of oil took a commercial turn with the increased usage of kerosene lamps, and new uses of kerosene were under development.[46]

Immediately after the Civil War, notable Americans such as John Rockefeller invested heavily in exploration and oil drilling through his Standard Oil Company. During the 1870s, oil became the major source of energy in the US, mainly due to the invention of oil-powered machines. Rockefeller's Standard Oil Company eventually controlled almost 90 per cent of US oil refining capacity and was a major distributor of petroleum.[47] The Company contributed to the development of petroleum as a major source of energy for the primary efforts of commercialisation.

Due to worries about the harmful effects of fossil fuels in 1860, a French innovator named Mauch Augustine developed the first solar power system. The invention aimed at harnessing solar heat for steam heat systems and drive industrial machinery. He aimed at developing a new clean and renewable energy source that could replace coal as a source of energy, as it caused a lot of pollution.[48]

Further advances in solar power utilisation came in 1876 with the development of the first solar cell. William Gryll and Richard Evans demonstrated that electricity could be generated by exposing selenium cells to sunlight.[49] The invention was later backed by famous innovators such as Warner Von Siemens, a contemporary of

Thomas Edison, who saw the potential of this new discovery. These discoveries initiated a lot of groundbreaking work in quantum physics and quantum mechanics, and more scientists were drawn to accepting and acknowledging the existence of atoms.[50]

Later in 1905 Albert Einstein published his theory of photoelectric effect that further reinforced the photovoltaic effects of light on metals. Einstein showed that light contains energy that is transformed when a packet of light that he called 'quanta' hits a metallic surface. Einstein boldly explained the phenomenon of light, and this gave scientists a new dimension to understanding the concept of transforming the sun's rays to electricity.[51]

The concept of photovoltaic effect was used to develop the first silicon solar cell in 1953 by the scientists at Bell Laboratories. Gerald Pearson, Darryl Chapin and Calvin Fuller of Bell Labs developed solar cells based on Einstein's ideas. The solar cells that they developed were able to generate measurable amounts of electricity using the sun as a source of energy. This marked a new era in solar energy use, and people were excited about the possibility of a limitless and renewable source of energy.

The potential of solar energy is evident with the postulation of massive solar cells that can generate electric energy using the sun's rays.[52] Commercial application of the silicon solar cell was boosted with the proposal by the US Air Force to use solar panels as a source of energy for the first US satellite. The proposed panels that integrated the use of chemical batteries and solar panels could provide power to fulfil the energy requirements of the satellites.[53]

Use of solar cells as an energy source became widespread in the 1970s with more homes installing solar devices to generate electricity to save cost on utility bills.

Initially, the use of solar power was a very expensive affair that many homes could not afford. However, in 1970 Dr Elliot Berman, in collaboration with the Exxon Corporation, devised new and cheaper solar cells that could be used in homes. The move saw a significant decrease in the cost of the solar cells, and this enabled more homes to have access to cheaper and cleaner lighting sources.[54]

The use of solar energy has a huge potential, and with the formation of the Solar Energy Institute and other relevant bodies to oversee the development of solar technologies there is hope for better applications of solar energy. It is projected that the world is likely to invent new generation solar-powered cars that can be cost saving and sustainable.

Electricity generation has undergone significant transformations since its inception as conceptualised by Michel and Jules Verne.[55] They viewed electricity as a reaction between the physical and chemical forces that caused vibrations of etheric particles. In the year 1882 Thomas Edison conceptualised the possibility of commercialising electricity for domestic and industrial use. Edison developed the commercial electric utility that supplied his customers in uptown New York with electricity.[56]

Later, in 1883, the first commercial use of hydroelectricity was developed in Appleton, Wisconsin. Installation of the electric power plants marked an important engineering milestone for both domestic and industrial users.[57]

The breakthrough in hydroelectricity was reached with the building of the Hoover Dam in 1935, which has the largest capacity to generate electricity. The hydroelectricity concept was later diversified to geothermal and nuclear power. Geothermal is considered renewable and, having the potential to contribute to the clean energy strategy, is being pushed by governments and other agencies. Sources of geothermal energy range from shallow ground to hot water rock found beneath the earth's surface.

In 1921 the first geothermal power station was privately developed by the owner of Geysers Resort in California which used a small turbine to generate electricity. This was followed by large-scale development of commercial geothermal stations in California that could generate a lot of electric energy.[58]

The most significant development in electricity generation came with the development of nuclear energy. Energy generated by nuclear power plants makes the bulk of electricity generated in developed nations. This is mostly due to the high capacity of nuclear reactors in generating electricity.[59] Today, however, there are moves and efforts by governments to limit the development and use of nuclear power, given the widespread environmental effects due to nuclear power plant disasters.

One of the most notable cases of nuclear disasters occurred on 28 March 1979 in a Pennsylvania nuclear reactor that resulted in the spread of nuclear dust over most of the nearby areas. Another similar event was reported on 26 April 1986 in Chernobyl Nuclear Power Plant in the city of Pripyat, Soviet Union.[60] The Chernobyl disaster led to widespread destruction in Russia and Ukraine, and caused many devastating effects on both the environment and the people.

The most recent nuclear accident to create a crisis situation happened in Fukushima Dai-Ichi Nuclear Power Plant on 11 March 2011, caused by a magnitude 9.0 earthquake. The accident resulted in the release of radioactive materials that triggered a 30-km evacuation zone. Due to the dangers posed by nuclear power plants, most governments today are looking for alternative and less hazardous forms of energy.[61]

In pursuit of clean energy the US government has instituted a series of initiatives to decrease carbon emissions by about 32 per cent by the year 2030. This will force industries to look for alternative, cleaner and more sustainable sources of energy. The government has further commissioned a sustainable approach towards reducing nuclear plants' carbon emissions by 20 per cent.[62]

There is a collective effort from all nations to shift energy systems from unsustainable to cleaner and sustainable sources of energy. Solar energy and wind energy have received particular focus, and more research is ongoing as to the best approaches to exploit these two sources of energy. The implication of the current move is the need to transform key sectors of the energy sector to meet current and future energy needs.

Notes

1 Maddison Historical Data. *World GDP, 1800–1900*. Accessed April 2016 from www.ggdc. net/maddison/oriindex.htm; more information on the Maddison Project can be found in the following link: www.ggdc.net/maddison/maddison-project/orihome.htm

2 A. Maddison (1995). *Monitoring the World Economy, 1820–1992*. Paris: Development Centre of the Organisation for Economic Co-operation and Development, 1 January, p. 255.

3 M. Roser (2016). *Economic Growth over the Long Run*. Accessed April 2016 from http://ourworldindata.org/gdp-growth-over-the-very-long-run

4 A. Maddison (2001). *The World Economy: A Millennial Perspective*. Paris: OECD, pp. 1–383.

5 General Etymology. *Population Dynamics*. Accessed April 2016 from www.cals.ncsu. edu/course/ent425/library/tutorials/ecology/popn_dyn.html

6 United Nations. *World Population, 1800–1940*. Accessed April 2016 from www.un.org/esa/population/publications/sixbillion/sixbilpart1.pdf

7 United Nations. *World Population, 1950–2015*. Accessed April 2015 from http://esa. un.org/unpd/wpp/Download/Standard/Population

8 The Maddison Project Historical Data (2013). *World GDP Evolution, 1820–1990*. Accessed April 2015 from www.ggdc.net/MADDISON/oriindex.htm

9 World Bank (2015). *World GDP Evolution, 1995–2014*. Accessed January 2016 from http://databank.worldbank.org/data/home.aspx

10 Maddison Historical Data. *World GDP, 1800–1990*. Accessed January 2015 from www.ggdc.net/MADDISON/oriindex.htm

11 World Bank. *World GDP, 1995–2014*. Accessed January 2016 from http://databank.worldbank.org/data/home.aspx

12 United Nations. *World Population, 1800–1940*. Accessed April 2015 from www.un.org/esa/population/publications/sixbillion/sixbilpart1.pdf

13 United Nations. *World Population, 1950–2015*. Accessed April 2015 from http://esa.un.org/unpd/wpp/Download/Standard/Population

14 M. Livi-Bacci (1992). *A Concise History of World Population*. Cambridge: Blackwell Publishers, pp. xvi–220.

15 World Bank. *World Total Population*. Accessed April 2016 from http://data.worldbank.org/indicator/SP.POP.TOTL

16 T. Malthus (1798). *An Essay of the Principle of Population*. London: J. Johnson, pp. i–125.

17 D. H. Meadows, D. L. Meadows, J. Randers and W. W. Behrens (1972). *The Limits to Growth* (1st edn). New York: Universe Books.

18 U. Bardi (2011). *The Limits to Growth Revisited*. New York: Springer Science & Business Media.

19 N. Wolchover (2011). *Will There Really Be 10 Billion People by 2100?* Accessed January 2015 from www.livescience.com/16479-10-billion-people-2100-population.html

20 DESA, UN (2015). *World Population Prospects: Key Findings and Advance Tables*. Accessed January 2016 from https://esa.un.org/unpd/wpp/publications/files/key_findings_wpp_2015.pdf

21 C. R. Coble, E. G. Murray and D. R. Rice (1987). *Earth Science*. Englewood Cliffs, NJ: Prentice-Hall, p. 102.

22 E. O. Wilson (2002). *The Future of Life*. New York: Vintage Books.

23 *The Travels of Marco Polo*, Book 1/Chapter 61, Of the City of Chandu, and the Kaan's Palace There. From Wikisource, translated by Henry Yule. Accessed January 2015 from http://en.wikisource.org/wiki/The_Travels_of_Marco_Polo/Book_1/Chapter_61

24 William F. Gericke (1940). *The Complete Guide to Soilless Gardening* (1st edn). London: Putnam, pp. 9–10, 38 and 84.

25 M. Verne and J. Verne (1889). Project Guttenberg in the year 2889. *Forum*, p. 262.

26 W. C. Otto (2015). Future energy system development depends on the past learning opportunities. *Wires Energy and Environment*, 19 (10), 172.

27 M. Verne and J. Verne (1889). Project Guttenberg in the year 2889. *Forum*, p. 262.

28 Ibid.

29 Ibid.

30 Ibid.

31 R. Heinberg (2005). *The Part is Over: Oil War and the Fate of Industrial Societies*. London: John Wiley.

32 IEA (2012). *Energy Technology Perspectives*. International Energy Agency, 16.

33 J. C. Williams (2006). *History of Energy*. Accessed April 2016 from www.fi.edu.com

34 L. Chen (2009). *China's Petroleum Industry*. Accessed April 2016 from www.worldenergysource.com

35 G. Ban (AD 32–92). *Hand Book Geography Annals*. Ji'an: Han Dynasty.

36 R. Wrighter (1996). *Wind Energy in America: A History*. New York: John Wiley & Sons.

37 L. Mumford (1934). *Technics and Civilisation*. Kinderdijk: www.mariajohannahoeve.nl.

38 A. Crosby (2006). *Children of the Sun: A History of Humanity's Unappeasable Appetite for Energy*. London: W. W. Norton.

39 NETL (2009). *History of U.S. Coal Use*. Accessed April 2016 from www.netl.doe.gov

40 NHA (2009). *The History of Hydrogen*. Accessed April 2016 from www.altenergymag.com/article/2009/04/the-history-of-hydrogen/555

41 NGSA (2009). *Natural Gas*. Accessed April 2016 from www.naturalgas.org

42 B. Kovarik (1998). Henry Ford, Charles Kettering and the fuel of the future. *Automotive History Review*, 32, 7–27.

43 Smithsonian National Museum of American History (2009). *Fuel Cell Origins: 1840–1890*. Accessed April 2016 from Americanhistory.si.edu
44 *Halladay's Revolutionary Windmill – Today in History: August 29*. Accessed February 2016 from http://connecticuthistory.org/halladays-revolutionary-windmill-today-in-history-august-29
45 *First American Oil Well*, American Oil & Gas Historical Society. Accessed April 2016 from http://aoghs.org/petroleum-pioneers/american-oil-history
46 U. B. Census (2010). *IHS Global Insight*. United Nations.
47 B. Black (2009). *Petroleum History: United States*. Accessed April 2016 from www.eoearth.org
48 J. Perlin (1999). *From Space to Earth: The Story of Solar Electricity*. Ann Arbor, MI: Aatec Publishing.
49 L. Mumford (1934). *Technics and Civilisation*. Kinderdijk: www.mariajohannahoeve.nl.
50 US Department of Energy, Energy Efficiency & Renewable Energy. *The History of Solar Energy*. Accessed June 2015 from https://www1.eere.energy.gov/solar/pdfs/solar_timeline.pdf
51 NETL (2009). *History of U.S. Coal Use*. Accessed April 2016 from www.netl.doe.gov
52 Southface (2009). *A Brief History of Solar Energy*. Accessed April 2016 from www.southface.org
53 NETL (2009). *History of U.S. Coal Use*. Accessed April 2016 from www.netl.doe.gov
54 J. Perlin (2009). *A History of Photovoltaics*. Accessed April 2016 from www.usc.edu
55 M. Verne and J. Verne (1889). Project Guttenberg in the year 2889. *Forum*, p. 262.
56 PBS (2009). *Edison's Miracle of Light*. Public Broadcasting Service. Accessed April 2016 from www.pbs.org/wgbh/americanexperience/films/light
57 ASME (2009). *Vulcun Street Power Plant*. American Society of Mechanical Engineers.
58 R. Wrighter (1996). *Wind Energy in America: A History*. New York: John Wiley & Sons.
59 NGSA (2009). *Natural Gas*. Accessed April 2016 from www.naturalgas.org
60 Z. Owen, M. Mackay, A. Ali, A. Sonia and B. Minsh (2015). *A 21st Century Power Partnership Thought Leadership Report*. NREL technical report. Denver: Clean Energy.
61 NPR (2011). *Timeline: A Nuclear Crisis Unfolds in Japan*. Accessed April 2016 from www.npr.org
62 M. Calleen and H. Army (2015). EPA emissions rules to mandate limits beyond proposed targets. *Wall Street Journal*. Accessed April 2016 from www.wsj.com/articles/epa-emissions-rule-to-mandate-limits-beyond-proposed-targets-1438488002

2 Scales and systems interactions across multiple energy domains

An energy system can be viewed as a value chain that consists of energy sources including wind, water, solar and oil to the consumers. The value chain helps in the distribution of energy to the end users by conversions.[1]

Multiple energy domains are tapped, after which conversion takes place and the energy transported through carriers to storage facilities or transported to the end users.[2] The main forms of energy include electricity and oil. Once the conversion takes place, energy is transported from the source of production to the point of consumption in the domestic, industrial or transport sector through a distribution system that consists of transmission grid systems, pipelines and other forms of transport.[3]

2.1 Scales and systems integration from residential to national

Integration at all scales across multiple energy domains is essential for efficient energy delivery. This is necessary because a lot of energy is lost as the energy passes through a value chain and reaches the consumers. Due to this, the supplied energy is lower than the converted energy. Optimum utilisation of energy therefore necessitates the need for scales and systems integration. In other words, there is a need for an integrated value chain for the efficient conversion of energy that is distributed to the end customers.

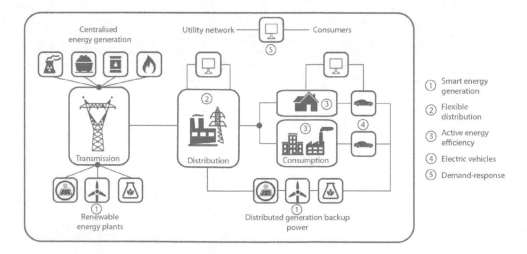

Figure 2.1 Modern energy systems and various facilities.

A modern transmission system consists of multiple value chains. Figure 2.1 depicts a typical modern energy system consisting of transmission, distribution and consumption with smart energy generation, flexible distribution and active energy efficiency.

Energy is tapped from multiple domains such as wind, water, nuclear and coal. Afterwards it is converted into electricity and distributed to the consumers.[4] An electricity system links different sources of energy with domestic and heavy industrial consumers, who use it to run appliances and machines.

The need for an integrated approach for dealing with energy demands has increased in the past few decades. A complex behavioural approach is conducted frequently for different activities at epistemological levels.[5] The complex approach in the urban areas entails interaction between structures, advancements in urban areas, and enhancement of systems optimisation modelling.

By contrast, in less developed areas the approach consists of communication of various systems from the physics of heat transfer right through the psychology as well as physiology of occupant behaviour and the socio-financial matters pertaining to the development industry.[6]

The present energy systems are developing through the expansion of interaction both in scale as well as multifaceted domain at progressive levels of interconnection. These levels consist of integration between electricity and gas systems due to the main role of electricity in controlling at multiple levels through the development in infrastructure of new energy systems.

Integration between dissimilar and diverse layers of energy systems occurs from a large number of measurements that include:

a) physical/technical (system equipment);
b) market and business (wholesale and retail, services and operations);
c) social (customers, users, stakeholders);
d) normative (administrative issues, guidelines and so on);
e) political (local, national, regional choice making and geopolitical ramifications);
f) digital (estimation, communication and control).

The combination of different distinct layers displays an increasingly multi-sided quality. This needs to be studied carefully to pinpoint the benefits of interconnections across space and time. One example is interconnected subsystems that link the wide transmission system with the neighbourhood distribution system.

Another example is the interaction between energy consumption and climate conditions in buildings through the operation of technologies.[7] However, despite the linked qualities of methodologies that are cited by the different research studies, a number of challenges are present for efficient integration at multiple levels.

Some of the challenges include spatial variables, accessibility and credibility of data, data intensity and particular information, integrated multilayer models, and policy relevance. Moreover, contingent upon the project and strategy chosen, information that is required might not be accessible. This includes the physical qualities of networks and buildings, verifiable energy utilisation of multiple activities, neighbourhood economic and macroeconomic indicators, as well as different climate information.

Whether the data is sourced simultaneously or freely, the raw information can vary in detail. The data can be gathered through surveys, open studies or examination from suppliers of energy. When developing the integrated models, one of the main challenges that most people will encounter relates to the interaction and dynamics between different spatial scales and capturing those dynamics at full and small scale levels.

Similarly it was observed that the current integrated model is strictly constrained to one spatial scale such as neighbourhood, building, city or district. In a real sense, energy consumption depends on different variables at different spatial scales that should be considered when upgrading infrastructure foundation in urban areas.

Energy system representation in current practice focuses on a main scale that shows a phenomenon at a macro scale with exogenous assumptions. For example, a scale from different points of view is an important factor used for forecasting energy usage after a certain point in time. Hourly load terms can change significantly and depend on the size of the appraisal. However, over a number of users the energy systems can carry on likewise without showing a solid smoothing impact by augmentation of the system boundary. During modelling systematic interaction, this methodology falls short in conveying a dynamic system model.

Energy storage options and control algorithms need to be considered in a unified manner in order to fully understand and manage energy systems. Moreover, it is also important to understand the spatial and temporal information between supply and demand infrastructures to fully grasp and manage such systems.

Integration of a scale and system is required to assess the energy system at different levels, from single buildings right through to neighbourhoods and cities. Additionally, the system model has to incorporate different sectors and not just explore the potentials of load management. A comprehensive model framework is required to fully understand the economic and technical impacts of taking on new energy policies and the latest technologies for different applications.

2.2 Scales and systems integration across multiple energy (and non-energy) domains

Energy–water–food nexus

Natural resources vary across different countries. Also, the pattern of the resource usage also varies across different regions. The actual usage of resources depends on accessibility as well as utilisation of materials from centralised to distributed coordination that is conducted under different legal agreements.

The complex nature of energy systems makes it difficult to monitor and interpret the energy resource nexus. This creates difficulty for decision makers to craft energy system policies.

To accurately assess the real system inter-linkages that are aligned with the global perspective, the simulation models should be integrated. This is required to ensure optimum representation of uncertainties during the modelling exercise.

The exact relationship between energy and water is complex and multifaceted. Water is required during different phases, including the extraction and processing of fossil fuel, growing biofuel or biomass, and for generation of electricity. In the same vein, a substantial amount of energy is required for treating, transporting and extracting drinking water and waste water.

In light of the above discussion, it is evident that no two resources can be considered separately. What is required is an integrated and holistic approach that will greatly help in grasping interdependencies between the resources.

Decision makers face confusion on how to support the nexus. This is mainly due to the immense size of each of the specific areas in the nexus system as well as the complexity of all three taken together, namely water, food and energy.

As a result, policies and regulations can unintentionally lead to suboptimal signals when it comes to the environment, economic concern and national security. Policymakers make the mistake of considering just two areas, with only a few approaches that can fully address the extensive interdependencies between the two systems. This is despite the fact that policy is structured by focusing on one area.

There is an increased need for systematic thinking during the decision-making process to overcome the complexity in the nexus.[8] Policymakers need to make a holistic policy and regulatory design while keeping in view such benefits as public health, increased options of livelihood, and resource and economic efficiency. Moreover, they also need to keep an eye on impact on the environment, commodity prices, infrastructure design, and society.[9]

Current modelling approaches

Development of national or regional energy policies is generally regarded as standing on three important pillars. These include environmental sustainability, economic competitiveness and energy security. This is particularly applicable in Europe, where each of the pillars is enhanced by a dedicated institution, including directorates for competition, environment, and transport and energy, respectively.

Development of effective energy policies is largely considered to be based on trade-offs between conflicting goals and strategies. In this context, various researches and simulation have been conducted together with the need for reducing conflict at geopolitical levels with regards to the energy nexus. The aim of the research study has been to ensure that the best policies are employed in the trade that provides optimum results.

A research study examined the climate–security nexus in the European energy sector.[10] The study utilised a model founded in scenario analysis that was created using the POLES model. It defined the results of GHG emission constraint in different degrees both at a global level as well as in situations that are specific for Europe, where there is an ambitious policy for addressing climate change unlike other continents that favour more self-effacing abatement policies. Findings of the research study demonstrated that drastic energy policy that addresses the climate issue can significantly benefit the respective countries included in the EU through two ways: lower susceptibility to price or supply shocks in the international energy market and climate-friendly energy policies.

Clingendael and the Energy Research Centre of the Netherlands (ECN) introduced a novel methodology and demand/supply index to measure supply security both in the long and medium term as well as a crises supply and aptitude index of 70 per cent for basic energy supply and 30 per cent for transportation and conversion.

A study that assessed the economic impacts of constraints in electrical supply in Slovenia during the period 2004 to 2008 introduced a technique that helped develop a composite index for assessing national electrical supply security as well as quantifying the risks.[11] The index was created from three dimensions that consisted of power system stability, primary energy supply security and environmental performance. These dimensions were further subdivided into other dimensions that were assessed by different indices. The sub-dimensions were then added into unique dimensional indices. Data employed were unique to the assessed electrical power system. Dimensional indices refer to contributions in the general composite index that are aggregated to a single composite electrical supply security index. That being said, the index did not allow direct comparisons or benchmarking between diverse systems.

The global policymakers need to resolve different interrelated complexities that have been characterised as basic threats to human civilisation.[12] Most of the issues relate to production, use and distribution of water, food and energy. And this is pertinent especially in developing nations.

Due to the immense scope of each of the areas as well as the increased complexity of each of the three resources taken together, very limited research work has been done to support the decision-making process relating to the nexus. As a result, policies and regulations unintentionally result in suboptimal results. To avoid this problem, there is a need for systematic thinking. This may not always bring results in the form of the legislative process of decision making, as focusing on all three or more resources in the nexus is complex and is bound to be influenced by one slight move made to exploit or utilise another resource in the nexus.

Effective policies apropos the food, water and energy nexus depend on the views of the policymakers. For instance, water and bio resources are considered as inputs, and resource requirements are viewed as outputs. Whatever the case may be, the perspective of the policymakers will significantly influence the policy that they propose for addressing the nexus. This is made possible by the priorities of a particular institution or ministry as well as the breadth and knowledge of the data analytical tools of the related support staff and experts.

Electricity and gas nexus

The electricity and gas nexus currently operates separately. By using different energy carriers, you must take a close look at the system that will empower interaction between them. When you understand the fundamental characteristics and qualities of each system, you will be in a better position to define the analogies. The fundamental terms that should be considered when managing the integrated electrical and gas power flow are: power (for the gas system it is pressure * flow (m^3/s), while for the electric system it is voltage * current (W)); flux (for the gas system it is flow (m^3/s), while for the electric system it is current (A)); and resistance (for the gas system it is friction factor, while for the electric system it is impedance).

Talking about representation of a network, a node in the gas network can either be a pressure node or a load node. Power requirements for load nodes are known at the start, and the only thing that needs to be resolved is the pressure value. These nodes are similar to a PQ load or load bus in the electrical systems. For a pressure node the potential is fixed, and they serve as a reference for other nodes. What needs to be calculated is flow injection while going through this type of load. They are similar to 'slack bus' or PV load in electrical systems.

Due to the increased usage of gas power plants in electricity generation in a number of countries, there has been a requirement to create models that would incorporate the extension and operational choices of electricity and gas networks. Some models have been established that analyse the interdependency between electrical and gas networks (see Figure 2.2). These models can be separated into two types: integrated electricity and gas network models, and energy system models.

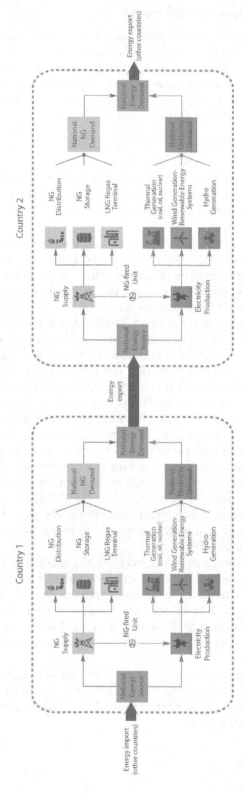

Figure 2.2 **Interdependency between gas and electricity networks and interconnection between countries.**

Integrated electricity and gas network models

Integrated models evaluate the interdependencies between gas and electric networks. The models achieve this aim by integrating electricity and gas models and using gas power plants as the physical connector. The integrated models make use of a point-by-point display of a few or a large part of the associated gas wells, gas storage, gas flow in the pipelines, gas demand, gas compressors, gas generation plants, electricity demand, and flow in the transmission lines (see Table 2.1).

Energy systems models

Energy systems models that have been depicted in Table 2.1 help in examining the interaction between different components of the energy systems, primary energy supplies, sectors of energy demand, conversion processes, and long-term planning and operation of the energy system.

Due to the complex nature of energy models, transmission and pipelines apropos gas and electricity networks, respectively, are modelled using transport mediums. These models look at the transport mediums before overlooking technical aspects such as pressure in the gas pipelines, gas flow rate, voltage levels, and frequency range relating to the gas and power flows in the system.

Infrastructures relating to storage are also simplified and sometimes excluded in the energy models. The author had reviewed the methodologies that were implemented by 37 energy models that utilised the integration of different renewable energy sources with different aspects of the energy system. The models range from local to global space limits in different time periods that are highlighted in this book.

The concept of 'energy hubs' is another way to analyse electricity and gas networks. The idea of the energy hub was first introduced in the Vision of the Future Energy Networks (VoFEN) programme. The objective of this project was to stimulate scenarios for the distribution and transmission of electricity in the next half a century, taking into account the flux in economy, nature and functionality.[13]

A number of papers have been presented up till now that shed light on the operations and management of the energy hubs.[14,15,16,17,18,19,20,21] Hydrogen as a carrier of energy has been considered as a main component of the energy hub.[22] The renewable part of the energy model as well as reliability of the energy systems have also been discussed at length by scholars.[23,24]

Energy hubs generally serve as a link between the manufacturers of energy, the energy transmission network and the end users. The hub can be characterised as consisting of different carriers of energy that are integrated with each other and make use of hub components including convertors, connectors and storages.

Size of the energy hub differs and can be represented as a country, a city, a local community or a building. The inputs and outputs of the energy hub can be integrated in a coupling matrix.

The most important component in the energy hub is the dispatch factor. Gas stream flow is partitioned into different parts (Figure 2.3). The first part flows into a CHP while the other parts flow into the heater. Rate of share of each of the convertors is represented by the dispatch factor.

Table 2.1 Review of methodologies on integrated electricity and gas network models

Applicability	Type of study	Model scope	Frequency	Time step	Year of publication (Reference)
National (Great Britain)	Optimisation model that reduces total operational costs under fault conditions to major gas pipelines to examine the impact of such faults to the integrated network.	Gas flow in pipelines, gas compressors, gas storage, generating power plants, transmission grid electricity demand and gas demand.	Monthly	Daily	2008[a]
IEEE 118 bus system	Security-constrained unit commitment (SCUC) model that minimises total operating costs under different fault conditions in the gas and electricity networks in order to analyse the interdependency between gas networks and security in the electricity network.	Gas network, gas storage, generating power plants and transmission grid.	1 day	Hourly and daily	2008[b]
National (Brazil)	Optimisation model that minimises investment and operational costs to assess the most cost effective expansion options in the gas pipelines, generation plants and transmission lines.	Gas wells, gas pipelines, gas storage, generating power plants, transmission grid, electricity load centres and gas load centres.	2009–2020	Hourly	2010[c]
National (Great Britain)	Combined Gas and Electricity Networks (CGEN) model minimises operational cost to determine the impact of large wind integration in the network on electricity generation profile and the gas network.	Gas flow in pipelines, gas compressors, gas storage, generating power plants, transmission grid, electricity demand and gas demand.	2 days	2 hours	2010[d]

(Continued)

Notes
a Modassar Chaudry, Nick Jenkins and Goran Strbac (2008). Multi-time period combined gas and electricity network optimisation. *Electric Power Systems Research*, 78 (7), 1265–1279.
b Tao Li, Mircea Eremia and Mohammad Shahidehpour (2008). Interdependency of natural gas network and power system security. *IEEE Transactions on Power Systems*, 23 (4), 1817–1824.
c Clodomiro Unsihuay-Vila, J. W. Marangon-Lima, A. C. Zambroni de Souza, Ignacio J. Perez-Arriaga and Pedro P. Balestrassi (2010). A model to long-term, multi-area, multistage, and integrated expansion planning of electricity and natural gas systems. *IEEE Transactions on Power Systems*, 25 (2), 1154–1168.
d Meysam Qadrdan, Modassar Chaudry, Jianzhong Wu, Nick Jenkins and Janaka Ekanayake (2010). Impact of a large penetration of wind generation on the GB gas network. *Energy Policy*, 38 (10), 5684–5695.

Table 2.1 (*continued*) Review of methodologies on integrated electricity and gas network models

Applicability	Type of study	Model scope	Frequency	Time step	Year of publication (Reference)
IEEE 118 bus system	SCUC model that considers the impact of constraints in the gas network on generation scheduling of gas power plants.	Gas flow in pipelines, gas compressors, generating power plants, transmission grid, electricity demand and gas demand.	Single time period	Hourly and daily	2011[c]
National (Belgium)	Integrated model that considers the effect of temperature in the gas network on the electricity flow and operating conditions in the model.	Gas flow in pipelines, gas compressors, temperature of gas in pipelines, gas storage, generating capacity of power plants, transmission grid, electricity demand and gas demand.	Single time period	Hourly	2012[f]
National (Belgium)	Optimisation model that considers the uncertainties of wind power generation on an integrated gas and electricity network in determining the optimum operating conditions of the system.	Gas flow in pipelines, gas compressors, gas storage, coal transportation system, wind turbines, pumped hydro power plants, transmission grid, electricity demand and gas demand.	Single time period	Single time period	2013[g]
National (Great Britain)	CGEN planning model minimises total operational costs and total system expansion costs to determine the electricity generation mix in a low carbon scenario.	Gas flow in pipelines, gas compressors, gas storage, generating power plants, transmission grid, electricity demand and gas demand.	2005–2030	5 years	2014[h]
National (Belgium)	Security-constrained optimisation model that minimises operational costs in order to determine stabilised states for the integrated gas–electricity network after simulating certain contingency scenarios.	Gas wells, gas pipelines, gas storage, generating power plants, transmission grid, electricity load centres and gas load centres.	Single time period	Seconds	2014[i]
IEEE 30 bus system	Integrated simulation model that analyses interdependency between gas and electricity networks during fault conditions and impact on security of interconnected network.	Gas flow in pipelines, gas compressors, gas demand, generating power plants, transmission grid and distribution network.	Single time period	Single time period	2014[j]

Regional (Europe)	Optimisation and simulation model for cross-border European trades to meet multidimensional requirements (economic, environment, security).	Gas flow in pipelines, gas compressors, gas storage, electricity and gas demand and gas turbine generators.	1 day	5 minutes	2015[k]
National (Iran)	Optimisation model that minimises total system expansion costs while integrating expansion planning of electricity generation, transmission and gas pipelines.	Gas flow in pipelines, gas compressors, generating power plants and transmission grid.	2, 3 and 6 year period planning between 2010 and 2016	Monthly	2014[l]
State (Victoria in Australia)	Optimisation model that maximises cost–benefit ratio in the integrated network when planning expansions to gas power plants in a low carbon electricity generation scenario.	Gas flow in pipelines, gas compressors, gas storage, gas power plants, transmission grid, electricity demand and gas demand.	1 day	Hourly	2015[m]

(Continued)

Notes

e Cong Liu, Mohammad Shahidehpour and Jianhui Wang (2011). Coordinated scheduling of electricity and natural gas infrastructures with a transient model for natural gas flow. *Chaos: An Interdisciplinary Journal of Nonlinear Science*, 21 (2), 25–102.

f Alberto Martínez-Mares and Claudio R. Fuerte-Esquivel (2012). A unified gas and power flow analysis in natural gas and electricity coupled networks. *IEEE Transactions on Power Systems*, 27 (4), 2156–2166.

g Alberto Martinez-Mares and Claudio R. Fuerte-Esquivel (2013). A robust optimization approach for the interdependency analysis of integrated energy systems considering wind power uncertainty. *IEEE Transactions on Power Systems*, 28 (4), 3964–3976.

h Modassar Chaudry, Nick Jenkins, Meysam Qadrdan and Jianzhong Wu (2014). Combined gas and electricity network expansion planning. *Applied Energy*, 113, 1171–1187.

i Carlos M. Correa-Posada and Pedro Sánchez-Martín (2014). Security-constrained optimal power and natural-gas flow. *IEEE Transactions on Power Systems*, 29 (4), 1780–1787.

j Burcin Cakir Erdener, Kwabena A. Pambour, Ricardo Bolado Lavin and Berna Dengiz (2014). An integrated simulation model for analysing electricity and gas systems. *International Journal of Electrical Power and Energy Systems*, 61, 410–420.

k Fatemeh Barati, Hossein Seifi, Mohammad Sadegh Sepasian, Abolfazl Nateghi, Miadreza Shafie-khah and João P. S. Catalão (2015). Multi-period integrated framework of generation, transmission, and natural gas grid expansion planning for large-scale systems. *IEEE Transactions on Power Systems*, 30 (5), 2527–2537.

l C. Spataru and J. W. Bialek (2014). *Energy Networks: A Modelling Framework for European Optimal Cross-Border Trades*. IEEE PES General Meeting, IEEE Xplore.

m Jing Qiu, Zhao Yang Dong, Jun Hua Zhao, Ke Meng, Yu Zheng and David J. Hill (2015). Low carbon oriented expansion planning of integrated gas and power systems. *IEEE Transactions on Power Systems*, 30 (2), 1035–1046.

Table 2.1 (continued) Review of methodologies on integrated electricity and gas network models

Applicability	Type of study	Model scope	Frequency	Time step	Year of publication (Reference)
State (Hainan province in China)	Optimisation model that minimises investment and production costs to determine which gas pipelines needed expansion to meet future energy demands.	Gas flow in pipelines, gas compressors, gas storage, generating power plants and transmission grid.	1 year	Minutes	2016[n]
National (Belgium), State (New England in United States)	Combined Electricity and Gas Expansion (CEGE) model minimises dispatch and upgrade costs in order to obtain the least cost extensions required in the integrated network under several cost and demand scenarios.	Gas flow in pipelines, gas compressors, gas storage, generating power plants, transmission grid, electricity demand and gas demand.	Single time period	Single time period	2016[o]

Notes

n Yuan Hu, Zhaohong Bie, Tao Ding and Yanling Lin (2016). An NSGA-II based multi-objective optimization for combined gas and electricity network expansion planning. *Applied Energy*, 167, 280–293.

o Conrado Borraz Sanchez, Russell Bent, Scott Backhaus, Seth Blumsack, Hassan Hijazi and Pascal van Hentenryck (2016). Convex optimization for joint expansion planning of natural gas and power systems. In *49th Hawaii International Conference on System Sciences (HICSS)*, pp. 2536–2545. IEEE.

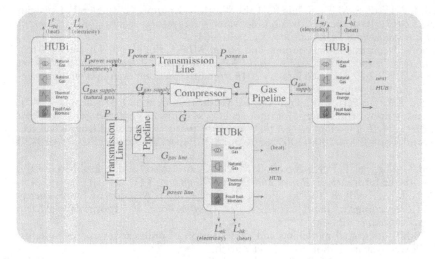

Figure 2.3 Energy hub: model of a local community energy system.

The energy system at the local level is composed of four different types of carriers of energy. These include the sun and wind that are oriented from the asset, and the electricity and gas that are oriented from the system. The four parts of the energy carriers match with two types of energy demand carriers, i.e. heat and electricity, via three hub parts that are categorised into five sub-parts: direct transmission (electrical transmission), convertors (PV panels, wind turbines and CHPs) and storage device (electrical battery).

Notes

1 E. Delyyanis and A. El-Nashar (2010). A short historical review of renewable energy. *Renewable Energy Systems and Desalination*, 1, 13. Accessed November 2016 from www.desware.net/Sample-Chapters/D06/D10-002.pdf
2 M. Hogan (2013). *Power Markets: Aligning Power Markets to Deliver Value. America's Power Plan.* Accessed from www.raponline.org/wp-content/uploads/2016/05/app-hogan-aligningmarketswithvalue-2013-sep.pdf
3 M. Roser (2016). *Economic Growth over the Long Run.* Accessed January 2016 from https://ourworld indata.org/Economic-growth-over-the-long-run
4 CERA (2013). *Energy Vision 2013: Energy Transition Past and Future.* London: CERA.
5 C. Spataru, A. Koch and P. Bouffaron (2015). Multi-scale, multi-dimensional modelling of future energy systems. *Handbook of Research on Social, Economic, and Environmental Sustainability in the Development of Smart Cities*, pp. 113–135. Hershey, PA: IGI Global.
6 Ibid.
7 Ibid.
8 C. A. Scott, S. A. Pierce, M. J. Pasqualetti, A. L. Jones, B. E. Montz and J. H. Hoover (2011). Policy and institutional dimensions of the water–energy nexus. *Energy Policy*, 39, 6622–6630.
9 M. Bazilian, H. Rogner, M. Howells, S. Hermann, D. Arent, D. Gielen, P. Steduto, A. Mueller, P. Komor, R. Tol and K. K. Yumkella (2011). Considering the energy, water and food nexus: Towards an integrated modelling approach. *Energy Policy*, 39, 7896–7906.
10 P. Criqui and S. Mima (2012). European climate–energy security nexus: A model based scenario analysis. *Energy Policy*, 41, 827–842.
11 Ibid.

12 R. Lawford, J. Bogardi, S. Marx, S. Jain, C. Pahl-Wostl, K. Knuppe, C. Ringler, F. Lansigan and F. Meza (2013). Basin perspectives on the water–energy–food security nexus. *Current Opinion in Environmental Sustainability*, 5 (6), 607–616.
13 United Nations. Department of Economic and Social Affairs, Population Division (2015). *World Population Prospects: The 2015 Revision, Key Findings and Advance Tables.* Working Paper No. ESA/P/WP.241. Accessed January 2016 from https://esa.un.org/unpd/wpp/publications/files/key_findings_wpp_2015.pdf
14 G. Andersson, P. Favre-Perrod, K. Frohlich, M. Geidl, B. Klockl and G. Koeppel (2007). The energy hub – a powerful concept for future energy systems. In *Proceedings of the Third Annual Carnegie Mellon Conference on the Electricity Industry.* Accessed from www.ece.cmu.edu/~tanddconf_2004/2007/2007%20Conf%20Papers/Andersson%20Paper%20final.pdf
15 M. Geidland, G. Koeppel, P. Favre-Perrod, B. Klockl, G. Andersson and K. Frohlich (2007). Energy hubs for the future. *IEEE Power and Energy Magazine*, 5 (1), 24–30.
16 P. Favre-Perrod, M. Geidl, B. Klockl and G. Koeppel (2005). A vision of future energy networks. In *Proceedings of Inaugural IEEE PES Conference and Exposition in Africa, Durban, South Africa*, pp. 24–30.
17 M. Geidl (2007). Integrated modelling and optimization of multicarrier energy systems. PhD thesis, Power Systems Laboratory, ETH Zurich.
18 M. Geidl and G. Andersson (2007). Optimal power flow of multiple energy carriers. *IEEE Trans. Power Systems*, 22 (1), 736–745.
19 G. Andersson, M. Geidl, K. Hemmes and J. Zachariah-Wolff (2007). Towards multi-source multi-product energy systems. *International Journal of Hydrogen Energy*, 32, 1332–1338.
20 L. Carradore and F. Bignucolo (2008). Distributed multi-generation and application of the energy hub concept in future networks. In *Proceedings of 43rd International Universities Power Engineering Conference, UPEC 2008, Padova, 1–4 Sept.*
21 M. Geidl and G. Andersson (2006). Operational and structural optimization of multi-carrier energy systems. *European Transactions on Electrical Power*, 16 (5), 463–477.
22 A. Hajimiragha, C. A. Canizares, M. Fowler, M. Geidl and G. Andersson (2007). Optimal energy flow of integrated energy systems with hydrogen economy considerations. In *Proceedings of IREP Symposium. Bulk Power System Dynamics and Control VII, Charleston, SC*, pp. 1–11.
23 M. Schulze, L. Friedrich and M. Gautschi (2008). Modeling and optimization of renewables: Applying the energy hub approach. In *Proceedings of IEEE International Conference on Sustainable Energy Technologies, Singapore.*
24 G. Koeppel (2007). Reliability considerations of future energy systems: Multi-carrier systems and the EECT of energy storage. PhD thesis, Power Systems Laboratory, ETH Zurich.

3 Whole energy system components

The energy mix consists of multiple parts of the energy systems supply that among other things consists of biomass that has been in use for 2 million years ever since the time when man first discovered fire.

Biomass long remained the main source of energy during ancient times. Humans used firewood for the longest period of time, mainly for lighting and heating.[1] It remained the primary source of energy before the discovery of fossil fuels.

During medieval times, most of the people used wood for heating, cooking and even in industries. However, after the 1800s, there was a shift to coal as the primary source of energy. The move towards coal was driven by the fact that it was more effective and efficient as an energy source as compared to wood. During the twentieth century, fuel oil accounted for 50 per cent of the total energy. Other components of the energy mix included oil, petroleum, electricity and natural gas. All these parts of the energy mix have been in use for decades. The only new component of the energy mix is nuclear power, whose share in the energy mix has increased considerably since the 1980s.

3.1 Overview of the whole energy system

At present, alternative renewable energy forms such as wind, photovoltaic and geothermal have been gaining ground as a prime source of the energy mix.[2] The world has seen a marked shift in the demands of alternative sources of energy, with a great amount of investment made in the development of these renewable sources of energy. During the past decades, the primary source of energy has transitioned from biomass to coal and later to oil (see Figure 3.1). The shift has occurred due to a number of factors. Some of the factors that have contributed to the shift in the energy source include efficiency and cost, availability, and suitability of the energy system for novel application and usages.[3]

Advancements in technology have fuelled in part the development of alternative sources of energy. Innovations and creativity are said to be important contributors to the shift from one energy source to another. Taking a look at the timelines will reveal that increased usage of coal was one of the factors that contributed to rapid industrialisation as well as leading to innovations such as the steam engine.

The discovery of oil led to the introduction of improved modes of transport systems.[4] It also led to the advancements in electricity that ushered in a new era of energy usage, opening up great potential and possibilities in different sectors of the economy.

Price and efficiency of energy sources have been major factors in the transition from one source to another. Future shifts in primary energy sources will depend mostly on the sustainability of energy sources and environmental impacts. At the moment,

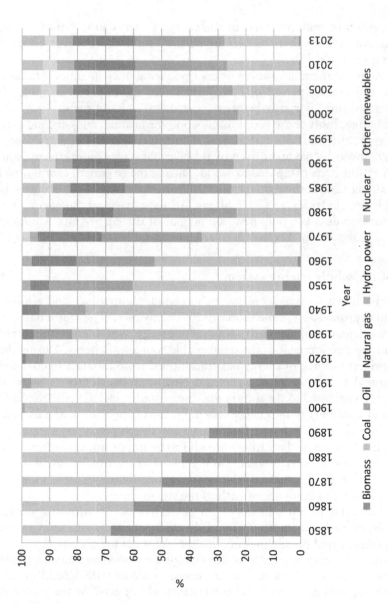

Figure 3.1 Share of fuels in the global energy mix across modern history. (Data source for 1850–1970[5] and for 1980–2013.[6])

the most pertinent concern as reflected in the World Energy Forum is the significant reduction of carbon emissions by the year 2020 through reducing our reliance on fossil fuels as the primary source of energy.[7]

At the World Summit on Global Warming 2015, a number of concerns were presented related to policy and political intervention to reduce the emissions of harmful carbon gases. The EU 2020 resolutions intend to introduce a carbon trading system. These efforts will set up a carbon trading system that is hoped will help in the shift to low carbon fuels, particularly hydrogen-based fuels.

The National Renewable Energy Laboratory has predicted that, in the coming five decades, about 80 per cent of the electric power supply in US homes will be sourced from renewable energy.[8] Similar predictions have been made in China's ambitious five-year plan that calls for a reduction in carbon emissions by making the shift to renewable, clean energy sources.

The world fossil fuel supply will be exhausted in the next one thousand years and the world will have to rely on alternative sources of energy.[9] At the moment around 87 per cent of the world's energy demands are being met with fossil fuels. That being said, given the present rate of usage of non-fossil fuel as a source of energy, the shift to alternative sources of energy will happen fast.

One of the major impediments to the shift to clean, renewable energy sources in present times is the abundant supply of fossil fuel as well as the reluctance to shift to alternative sources of energy, mainly due to cost.[10]

Another major obstacle in the shift to alternative energy sources is the decisions made by policymakers in respective countries. Note that the predictions about energy usage that have been depicted in Figure 3.2 largely depend on the policies that are developed now relating to the development of new energy sources. This may take a lot of time, similar to the transition that occurred in the past apropos the energy mix.

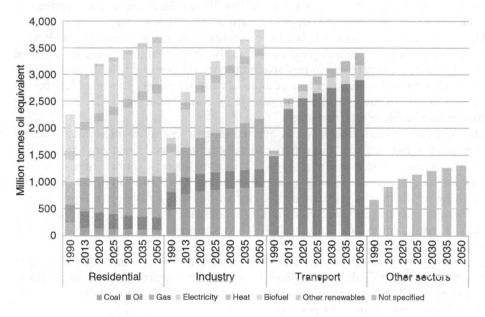

Figure 3.2 Energy demand per sector. Data source: 11

3.2 Energy demand

Population plays a pivotal role in the demand for energy. As described in Chapter 1 of this book, population has increased significantly in the past few centuries. Increase in population is an important factor that contributes to the change in energy mix.

Currently the demand for fuel oil dominates other energy sources, mainly for use in industries and transport. The share of oil as an energy source stands at the moment at 1.5 per cent, while use of coal and biomass stands at 47 per cent and 47 per cent, respectively. The increase in private car ownership has been one of the main factors resulting in increased global demand for fuel oils.[12]

The rising demand for oil has been spurred mainly by increased oil supply. The glut in the oil supply has made oil cheaper in the international market. That has been one of the contributing factors for the increase in the demand for oil.

The oil price embargo that the OPEC member countries orchestrated during the 1970s created reverberations in different sectors of the economy. This raised the need to discover alternative energy sources.[13] Disruptions in energy due to the oil embargo led to energy insecurity, due to which a number of countries decided to switch to alternative sources of energy.[14] This was also one of the factors that led to the development of nuclear energy and biofuel to produce electricity.[15] However, at the same time, increased usage of transport fuelled the increase in demand for fossil fuels that today accounts for 87 per cent of transport energy.

Use of natural gas has increased in the past decades due to the discovery of new industrial usages of the resources. The natural resource is viewed as an alternative to coal and oil. With the globalisation of energy, the demand for natural gas is likely to increase. The fact is that natural gas has found use in both the domestic and transport sector (see Figure 3.2). However, transport of gas from the point of origin to consumption requires installation of costly infrastructure that increases the price of fuel charged to the consumers.[16] Companies such as BP and Shell are at the forefront of tapping natural gas reserves and supplying natural gas to consumers.

Electricity generation is one of the major outputs of energy systems. Use of electricity as an energy source had increased considerably after its commercialisation during the late nineteenth century. Countries around the world invested heavily in electricity as it was cleaner and easier to use as compared to other energy sources. New inventions such as accumulators and dynamos in 1888 contributed to the development of electricity as a viable source of energy.[17]

The invention of machinery that converted electric energy to mechanical energy such as electric motors, fans and others radically transformed the energy sector. Use of electricity as a source of energy will continue to increase in the future, with a greater number of industries using powerful machinery that runs on electricity.[18]

By the year 2025, it is postulated that electric vehicles will be in widespread use in almost every major city in the world. Global demand for electricity will continue to increase in the future, fuelled mainly by the increase in population as well as the development of the economy.

According to a report from the U.S. Department of Energy, the share of coal in electricity production in the country has decreased from nearly 48 per cent to about 36 per cent. In China, coal is used in the generation of 80 per cent of the electricity, while in Europe the figure is 31 per cent.

Domestic energy demand modelling: a theoretical background

Energy demand modelling requires using simulation software. There are two basic kinds of simulation systems: dynamic and non-dynamic. Dynamic models are fairly complex with moderate-to-slow simulation speed and work on hourly inputs with detailed descriptions that incorporate all the interacting dynamics and provide detailed answers. Non-dynamic is a basic system with quick simulation speed that works on monthly and yearly inputs and outputs with non-detailed answers.

Whatever the model that is used to predict energy demand, there are a number of obstacles that present difficulty in the accurate prediction of energy demand. One challenge in simulating demand is the sheer amount of the energy demand users. For example, in the UK there are more than 60 million individuals and 25 million homes. Simulating energy demands at such a scale presents many challenges, including chances of errors, speed of execution, and accuracy of the results.

Forecasting energy demands using simulation software is susceptible to human errors while inputting the data. The house environment system is chaotic and sensitive to initial conditions, has non-linear dynamics and is deterministic. In other words, individual houses have different energy demands during different points of time. This creates difficulty in accurately predicting the energy demand.

The demand accuracy gets worse when making estimations further in the future – a phenomenon that is called the butterfly effect. This calls for the need to make assumptions about what drives household energy demand. For instance, population service requirements in homes drive energy demand. Understanding the characteristics of the population will help increase the accuracy of the energy demand prediction.

3.3 Energy supply

Demand for energy has increased gradually with the advancement of technology and the development of industries. The overall consumption of energy in the world has increased by about 27 per cent since the beginning of the new millennium. The rate of energy consumption of both renewable and non-renewable energy sources is increasing continuously.[19]

Increase in demand for energy has resulted in greater need to secure a reliable supply of energy. Recently, renewable sources of energy have made major contributions in meeting the demand of the consumers. Renewable supply sources including geothermal, hydroelectric and solar systems provide 4 per cent of the global energy demand. And this figure will increase in the future as more countries invest in alternative renewable sources including sun and wind.[20] In certain areas whole communities have developed that depend on renewable energy sources to meet energy demands.

Biofuel is another renewable source of energy that is being pursued in different countries. Global supply of the renewable energy source grew by nearly 6 per cent with more than 1.9 million barrels of biofuels being produced at a commercial scale.

Nuclear fuel represents 12 per cent of energy that is produced globally.[21] Most countries resorted to nuclear energy to get rid of their reliance on coal and oil as a primary source of energy. Although nuclear fuel is less environmentally harmful than either coal or oil, the risk of a nuclear disaster makes it equally harmful as the other two modes of energy. As a result, increasing numbers of countries are resorting to alternative sources of energy supply to meet the demand. That being said, at the moment nuclear energy is the only option for a number of industrialised countries for generating electricity that does not result in carbon emissions.

The International Energy Agency (IEA) report shows that natural gas is becoming a popular source in many countries. By the year 2020, it is expected that natural gas will comprise about 50 per cent of the total used to produce electricity.[22] Almost half of the world's oil and gas supplies comes from the Middle East, which still remains a dominant supplier of fuel oil and gas in the world.

Figure 3.3 shows the increasing trade in natural gas, and there is evidence of a growing supply of oil (see Figure 3.4) and gas with new oil and gas reservoirs. An observation by John Watson indicates that the new technologies are assisting in the exploration and discovery of more oil and gas from oil fields that were previously impossible (May 2014). For example, more countries are initiating offshore drilling, and shaling and fracking are making it possible to exploit oil from different oil fields. The supplies from Canada stand at 1.7 million barrels per day, and this is expected to increase with the offshore drilling activities.

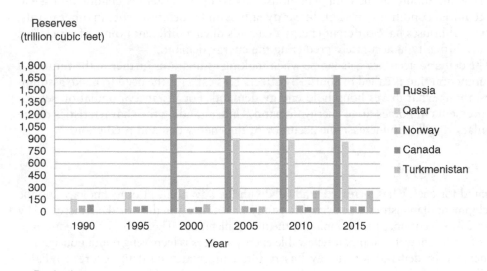

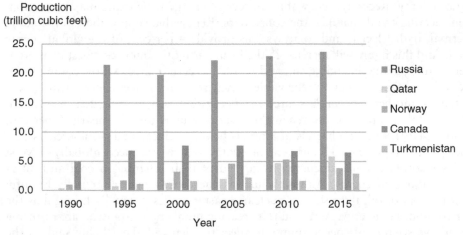

Figure 3.3 World natural gas top five countries for reserves and production. Data source: 23

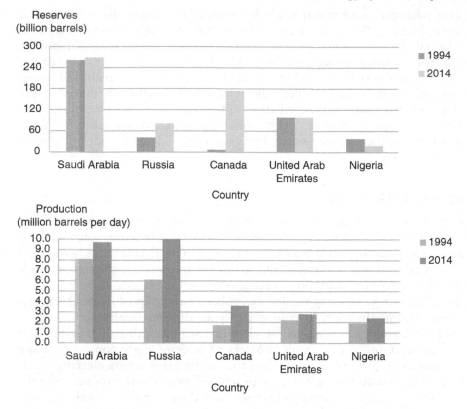

Figure 3.4 Top five countries for crude oil reserves and production in the world. ^{Data source: 24}

3.4 Energy storage

Energy storage need in the future depends on different factors such as fluctuating demand and supply, the energy sources mix, capability of shifting demand, specific energy utilisation and interdependencies between multiple energy vectors.

Storage of energy entails costs as well as energy losses. That's why it is utilised only when there is a significant gap between energy demand and supply.

A number of obstacles are present in setting up energy storage infrastructure. The first challenge includes doubt about the grid level in which electrical systems must be set up.[25] A way to work round this problem is to integrate privately operated or smaller systems in the distribution network.

Economic deployment of electrical storage systems depends on favourable policies. The presence of ineffectual policies serves as an obstacle to setting up energy storage facilities. The third challenge is developing business models that will help in the effective deployment of storage systems.[26]

Different studies in the past have shown that efficient storage of electricity can result in savings in the grid systems. However, gathering cumulative benefits and developing a viable proposition is not an easy task.[27]

Most of the current electrical storage systems (e.g. UK) are in the form of pumped hydro storage (PHS).[28] This type of storage facility is used by power generation companies in load balancing. During low demand, energy is stored in the form of the gravitational potential energy of water, which is pumped to a higher elevation reservoir from a lower location. When the electric demand becomes high, the stored water is released through the turbines that generate electricity to meet the fluctuating demand.

A much more superior storage solution to PHS is the distributed storage method (DSM). The storage system is ideal for isolated networks, smart grids and distribution network reinforcement.[29] The advantage of this method is that they are flexible technologies that are situated close to the load. The socio-technical shift of DSM into future power networks will probably provide greater value.[30] Despite the clear benefits of the storage system, however, it has not been widely deployed for electricity storage.

Modelling tools and techniques

A lot of scholarly contributions have been made apropos energy system modelling. This is despite the fact that models have been used relatively recently to represent electrical storage systems. That being said, representing electrical systems in the form of models is a difficult task.[31] The difficulty arises due to short operational timescale as well as simultaneous maximisation of energy storage and release.

Another difficulty in utilising modelling tools and techniques relates to aggregating value streams that will help in quantifying the electrical system value.[32] Two broad methodologies of modelling electrical storage value include market-based and production cost methods.[33] Market-based methods consist of using historical data to evaluate or estimate the value of the price-taking electrical storage systems. This assumption justifies the practicability of the technique, since the operation of the electrical storage system will not disturb previous values. Some scholars relax this assumption by examining the effects of saturation of the market on peak prices.[34]

Using models to stimulate energy storage solutions can offer deep insight into the pathways of ES deployment. Pathway analysis is made possible by utilisation of dynamic optimisation models. Two examples of an effective model include the TIMES Model and the ETI ESME Model.[35,36] These models are used to update policy-makers about demand and supply through simplistic simulation of electrical systems. The holistic technique can be extended to model bulk PHP systems exclusively. A shortcoming with these models is that they usually have five-year time-steps that in some cases makes them inappropriate for electrical storage stimulation.[37]

Energy system models can offer initial conditions that can be used in other models. Using the two models together allows detailed analysis and predictions of energy systems. This approach is used by the ETI that utilised PLEXOS and ESME simultaneously to gain deep insights into the electrical systems.[38]

An effective model that can stimulate energy systems within a long-run framework is Digital Station Intelligence Manager (DSIM).[39] This method was developed by scholars at Imperial College London and is cost effective, reliable and, most importantly, power efficient.[40] The DSIM model can stimulate the electrical storage system while utilising minimum costs. This is made possible through considering generation and investment costs, reserve services and wind curtailment.

Two other market-based analyses that make use of customised econometric methods or spreadsheet-based techniques include ES-Select™ and ESVT.[41,42] The former is publicly available and can be accessed from the Sandia National Laboratories, while the latter is commercially available. There are many other production costs tools that can be used to stimulate electrical storage systems including PROSYM, PROMOD and PLEXOS.[43]

In the grid-level hierarchy, operators at the top hierarchical levels are able to handle more energy, serve greater geographical areas, and operate larger systems. They are also more controlled compared to operators at the base levels. Those at the base-level areas are relatively more distributed, numerous and make use of small systems.

Transmission system operators are the top-level operators that make use of bulk electrical systems including CAES and PHS. They offer a number of benefits to the grid energy system including societal benefits that cannot be monetised.[44,45] A number of studies have examined electrical storage at grid scale. Utilising the ESVT model at grid stations in California, it was found that 50MW batteries can be effective in serving as bulk storage and offer ancillary services.[46] Another study used PLEXOS and found that electrical storage offers net savings on the island of Maui.[47]

Additionally, NaS and Li-ion batteries were found to be relatively more effective as compared to natural gas in offering support to 15GW of estimated wind power in the Northwest Power Pool (NWPP).[48] Stakeholders most often support use of a grid-scale electrical system to allow for the expansive penetration of energy systems.

A UK-based study has utilised the DSIM model to show that electrical storage can result in savings of about £1.8 billion.[49] Although the study did not mention anything about value stream aggregation, it did find that DSR and integration do not considerably affect the electrical storage value.

In contrast, the system found synergy between the two. The study also provided evidence that electrical storage offers more benefits in terms of renewable-oriented pathways that rival nuclear pathways. Lastly, the long-term benefits of distribution storage over bulk storage are found mainly to be due to reduced distribution infrastructure costs. According to an estimate, electrical storage has the potential to provide savings of £2–10 billion in the next three decades.[50]

Grünewald *et al.* (2011) evaluated the potential of grid-scale electrical storage by utilising demand profiles as well as the previous year's meteorological data as inputs along with a high temporal resolution.[51] Their study found that the system can only be profitable in the case of wind turbine usage. This will result in cost reductions and trading margins.

Distribution Network Operators (DNOs) and substations comprise the next two hierarchical levels. Electrical storage at this level is more distributed and relatively more common at the substation level. It is estimated that around 100,000 batteries are used to offer onsite power at the substation level in the US.[52] These batteries are not practical for use in other scenarios. However, there is great use of the batteries at the distribution and substation levels.

Two contractual structures were suggested that helped in optimising the exposure of risk as well as technical capabilities in the current regulations. One is the DNO model that viewed electrical storage systems constructed by the DNOs that are used in particular security windows and contracted to external parties. The other model is the contracted service model in which the DNO contracts an external part electrical storage system provider for utilisation in specific windows. A third model was also proposed but was rejected due to regulatory restrictions.

The above study was not in any way quantitative in nature. Later studies highlighted that the grid-scale electrical storage system can offer distribution-level services while remaining profitable.[53]

Another study described the successful use of 1.2MW NaS batteries to offer peak shaving as well as transmission upgrade deferral services in the state of West Virginia.[54] The electrical storage system was found to be more costly as compared to traditional solutions. However, the benefit of such a system is that it can be installed quickly and flexibly deployed in other locations, incurring long-run cost savings.

The lowest level in the hierarchy of grid levels is occupied by distributed operators. At this point, electrical storage is mostly used for offering operational support as well as seeking profits. A number of literature studies have focused on this level of the grid. The economics of the distributed electrical storage system was analysed in a study using past price data.[55] Findings of the study reflect that flywheels can be operated utilising reduced costs. However, the NaS system cannot be used effectively for electrical storage. This means that power quality functions offer more value as compared to reserve services and arbitrage.

Other studies have extensively evaluated the renewable-oriented use of electrical storage as well as a market modelling approach to find profitability in the system.[56,57]

Most of the UK-based studies focused on the use of a distributed electrical storage system for supporting renewables. A study used linear programming optimisation to find the maximum dispatch algorithm to offer renewable capacity firming.[58] The optimal model was found to be one where the electrical distribution system consisted of two wind farms that resulted in optimal profitability.

The value proposition of an electrical storage system is found to be important in performing renewable capacity firming and arbitrage in the UK energy market.[59] Electrical storage systems that are high power and short duration will most likely be profitable as compared to high-capacity systems that are uneconomical.[60] The storage system services can be worth in the range of £500–800/kW per annum on the assumption of 26GW wind power.[61]

Electrical storage systems have been found to offer savings of between £40/kW and £125/kW depending on the wind penetration.[62] Moreover, there is a considerable synergy between the electrical storage systems and wind turbines that increases the value of the energy system.[63]

Current and anticipated technologies

In Table 3.1a and b you can find important characteristics of present and estimated electrical storage systems. Fuel cells have been omitted as they are expensive to implement at the present time. Information presented in the table is compiled from different study sources.[64,65,66,67,68,69,70,71,72]

The ranges given in Table 3.1 show that there is some vagueness about the technical capabilities of the energy storage system, either due to immaturity or other reasons. Capacity can either be reported in units of energy or time, depending on what is the more appropriate for the situation.

From the same sources that were used to compile information for Table 3.1b, information on the practical and economic aspects of the technology were collected. Lifetime technology is reported in years, cycles of which depend on the application that it will serve. For instance, an operator might put more value on a flywheel compared to its lifetime.

Table 3.1 ESS technical characteristics

Table 3.1a

	Mechanical			Thermal		Electrical	
	Flywheels	CAES	PHS	CES	PHEES	SMES	Supercapacitors
Power (MW)	0.1–1	50–400	10–40,000	0.25–10	2–5	High	High
Capacity	10–100s	8–26 hrs	500–8,000MWh	Scalable	Scalable	Low	Low
Power density	High	Low	Low	Very low	Very low	High	High
Energy density	Low	Low	Low	High	High	Low	Low
Efficiency (%)	85–93	60–80	65–85	40–70	72–80	90–97	75–80
Response time	Seconds	Seconds	Around a minute	Minutes	Minutes	Milliseconds	Milliseconds
Self-discharge	Very high	Very low/none	Very low/none	Low	Medium–low	Very high	Very high

Table 3.1b

	Batteries					
	Li-Ion	NaS	NI-Cd	Pba	VRB	Zn-Br
Power (MW)	1–100	1–35	45	10–50	0.25–12	0.1–1
Capacity	50MWh	10–245MWh	6.75MWh	40–300MWh	2–120MWh	0.1–4MWh
Power density	Medium–high	Low	Low	Low	Low	Low
Energy density	High	High	High	Low	Low	Low
Efficiency (%)	87–92	75	75	63–90	65–88	65–85
Response time	Milliseconds	Milliseconds	Milliseconds	Milliseconds	Milliseconds	Milliseconds
Self-discharge	Medium–low	High	Low	Low	None	None

For mechanical storage systems, CAES requires high capital cost, it is geographically limited and it can require natural gas for operation, while flywheels have a medium to low cost and some have already been deployed. In the case of electrical storage systems, supercapacitors and SMES have both high cost and good dynamic performance. However, both are at the demonstration stage. Thermal storage systems, such as CES and PHEES, can use waste heat for generation, which makes them very sustainable options.

Energy storage worldwide

Energy storage is not a new concept related to electricity systems. The method is an important component of the storage system that is used for managing load. Various studies have examined the development of energy storage in leading industrialised nations including the US, Germany and Japan.[73,74] The studies have discussed current advancements as well as drawn lessons that will help in pushing forward energy storage systems.

Challenges to the organisation and advancement of alternative technologies of energy storage can be gathered into three main categories: technological, market and regulatory, and strategic.

A number of obstacles are present at the technological levels. Some of the obstacles include the following:

a) *Capacity*. There is a requirement to enhance the power and optimise the capacity limit through efficient usage of the existing technologies. This requires advancement of new materials and processes. However, innovation in new batteries can take about two decades to move to a commercial centre from the research facility.
b) *Practicality*. Another important factor is the practicality of deploying the electrical storage facility at the small and decentralised stage. This is required for market penetration that requires the introduction of advanced technologies. At the moment advancements are being made in the form of deployment of packaged battery units in the market for domestic use.
c) *Cost*. The third obstacle is the cost of the system that should be addressed carefully before deploying the system. This can be achieved in multiple ways including learning by doing, economies of scale, and further development.
d) *Market and regulatory issues*. A number of market and regulatory issues also present a problem in the deployment of energy storage systems at a technological level. Relevant market signals need to be present that should act as an incentive for building storage services and expanding storage capacity. This may include providing support in the short term that assists the new participants to compete effectively and benefit from economies of scale.

Overcoming technological barriers

There is a great disparity between nations when it comes to development of the energy storage system. The Japanese government has taken an approach that is best described as 'battery driven'. The approach has been very practical in that it focuses on a particular execution upgrade. When it comes to the US, the approach is less innovation driven and more focused, mainly in communicating advancements of the phases to the investors. Germany, on the other hand, has taken an innovative approach with a focus on capacity innovation to meet temporary points, while at the same time work is in progress to fulfil a more drawn-out objective.

Market and regulatory developments

The main thrust of the regulatory framework is to create a level playing field that allows trading in cross-border electrical storage. This necessitates the need of a regulatory framework to offer clear obligations and principles relating to the technical and financial conditions.

The choice to utilise resources for the improvement of storage capacity and adequate capacity will depend to a large extent on the advancement of the whole system. This is closely linked to a number of advancements such as electricity super-highways together with large-scale arrangements of renewable energy generation especially in North Africa and the North Sea. There is also an increased need for development of improvements in demand side management and the development of electric vehicles.

3.5 Super grids

A super grid is a kind of electricity transmission system that is based on high voltage direct current (HVDC). The system is suitable for facilitating large-scale and sustainable power generation over a long distance.

A number of factors contribute to the need to create super grids. These factors include the following:

- 10 per cent loss incurred in the case of transmission distance over 3,000km in comparison to 40 per cent for HVAC lines;
- Lack of technical limit to potential length since no reactive losses are present in HVAC lines;
- Interconnect AC networks that operate at multiple frequency;
- Less impact on the environment when using two lines as compared to using lines in HVAC for the same amount of electricity;
- Fewer protection devices due to low short circuit current;
- Lower investment cost of HVDC over a distance of 800km on land and 50km at sea; and
- Better security and sustainability of carbon neutral, supply and single electrical markets with super grids.

Let us now take a look at some of the super grid projects around the world.

Super grid projects around the world

a) North Seas Countries Offshore Grid Initiative

The North Seas Countries Offshore Grid Initiative was introduced in 2008 by the European Commission. The aim of this project was to develop offshore wind in the North Sea, Baltic Sea and Irish Sea, with the first project in the North Sea. A Memorandum of Understanding was signed by ten countries in 2010 – the UK, Germany, France, Belgium, the Netherlands, Luxembourg, Denmark, Sweden, Norway and Ireland.

The project aimed to contribute to the shift to a sustainable low-carbon economy while at the same time maintaining security of the energy supply in a cost-efficient manner. It was projected to connect 55GW offshore wind capacity using 10,000km (HVDC and HVAC) transmission lines by 2030.

b) DESERTEC (Energy from the Desert)

The DESERTEC programme was launched in 2003 by the Club of Rome – a global think tank organisation. Article 9 of DIRECTIVE 2009/28/EC of the European Parliament and of the Council on the encouragement of the utilisation of energy from renewable sources governs how renewable electricity could be imported from non-EU countries to the EU.

About 12 EU companies had joined the programme in 2009 with the objective of supplying 17 per cent of EU electricity demand from renewable energy sources in MENA (Middle East and North Africa) by 2050.

Electricity generated from CSP plants in MENA can provide a reliable and controllable base power to load centres in the EU and also balance power deficits through HVDC transmission lines – the undersea HVDC transmission line infrastructures will need to be upgraded to successfully transmit renewable energy imports from North Africa to large load centres in the EU and MENA.

An interconnected EUMENA grid and market based on renewable energy has potential benefits for both regions, such as considerable reduction of CO_2 emissions from the power industry in EUMENA, less costs to the entire EUMENA compared to individual countries implementing renewable integration, and economic development for countries in MENA through a large-scale export electricity industry.

There are a number of challenges faced by the DESERTEC project. The project is too expensive – it requires an additional investment of €400 billion up to 2050. Also, 47 out of 50 shareholders have left the consortium, as the project was not seen to be viable to them in terms of return on investment. Moreover, the EU transmission grid is currently struggling to successfully integrate the additional renewable energy generated in Europe. Lastly, resistance from MENA countries also undermines the success of the project due to concern about the benefits to their respective countries. Imports from MENA will require significant changes to the current technical, economic and regulatory framework.

c) MedGrid (Mediterranean Grid)

MedGrid is a promising initiative that was launched in 2008 by the French government. The objective of the project was to generate 20GW of electricity from renewable plants in North Africa with 5GW exported to Europe by 2020.

The initiative is similar to the DESERTEC programme but with a focus on concentrated solar power (CSP) in North African countries. Under the MedGrid programme, electricity will be transported to North Africa using HVDC lines that cross the Mediterranean Sea. The investment cost of the project amounts to about €5 billion.

In 2011, DESERTEC and MedGrid signed an agreement to coordinate their activities on the development of renewable energy from the deserts and suitable transmission infrastructures. A 160MW CSP plant in Morocco was completed in 2015, with more CSP plants under construction in Morocco, Egypt and Algeria.

d) West Africa Power Pool and current situation

West Africa (WA) Power Pool was formed in 2000. The project consists of 14 countries (Benin, Burkina Faso, Côte D'Ivoire, Gambia, Ghana, Guinea, Guinea Bissau, Liberia, Mali, Niger, Nigeria, Senegal, Sierra Leone and Togo). It has been estimated that it will introduce 10GW of hydro and gas generation plants as well as link the 14 African countries through 7,000km of transmission lines by the year 2019.

There is great potential for hydropower resources to meet 60 per cent of the energy demand in the WA region. Major importers of electricity in the western part of the continent include Benin, Burkina Faso, Nigeria and Togo.

The electricity trade between Western African nations is limited as generation plants and transmission lines are in construction phases. For some countries such as Côte D'Ivoire, Niger and Nigeria that are facing difficulty in meeting their own energy demands, this initiative may result in relatively cheap electricity generation from gas and hydro.

e) Gobitec and Russia's hydropower

Gobitec and Russia's hydropower super grid project was proposed in 2009 to source energy from renewable sources in the Gobi desert as well as large hydro resources in Russia. The grid will deliver electricity through an HVDC grid that connects Russia, Mongolia, China, South Korea and Japan. The estimated cost of the project is $293 billion, and by the year 2030 the power plant that will source energy from solar and wind plants will expect to have a capacity of 500MW.

The project comprises three stages. The first stage lasts three years and consists of building a 50km length of HVDC lines, wind power plants with a capacity of 100MW, and gas plants with a capacity of 1GW.[75]

The second stage will last for four years and entails the construction of a 2,300km length of HVDC lines, wind power plants with a capacity of 1GW, gas plants with a capacity of 0.6GW, hydropower plants with a capacity of 1GW, and solar power plants with a capacity of 0.5GW.

The third stage will last for about five years. This stage will consist of building a 4,000km length of HVDC lines, wind power plants with a capacity of 1.5GW, gas plants with a capacity of 1GW, hydropower plants with a capacity of 0.6GW, and solar power plants with a capacity of 0.5GW.

f) Asia–Pacific SuperGrid

The Asia–Pacific SuperGrid will connect Northern Australia to four countries in Southeast Asia, namely Timor-Leste, Indonesia, Malaysia and Singapore. The super grid may also be used to connect Indonesia's geothermal resources to countries in the region.

When completed, the Australia–Asia Interconnector will be one of the deepest and longest sub-sea HVDC cables in the world, having a length of 700km and depth of 3km. A future initiative consisting of generating 1,000MW capacity from solar power in Northern Australia by 2050 that has an expected investment of $10 billion is also being investigated.

g) IceLink

The IceLink project is a proposed connection of a 1,000km cable between Iceland and the UK. When the project is completed, it will be the longest sub-sea interconnector in the world that will bring 1.2GW of sustainable power from hydro plants.

h) Brazilian Super Grid

This is a 2,385km transmission grid (HVDC and HVAC) that will connect a 10GW hydro power plant in the Amazon Basin in the north-west to populated areas in south-eastern Brazil.

i) Atlantic Wind Connection

Atlantic Wind Connection is a proposed 560km HVDC transmission grid located on the east shore of the United States. It will connect off-shore wind farms to New Jersey, Maryland, Delaware and Virginia.

Global super grid – a proposition

A global super grid that connects power plants in the world using HVDC transmission lines can offer various benefits. These include a secure energy supply through well-defined agreements, balanced demand and variable supply, reduced power reserve and stabilised electricity prices through a single market.

Such a global grid will consist of offshore wind plants, solar power plants in deserts, large hydropower plants in Greenland and a few fossil fuel plants. The proposed global super grid could result in a considerable reduction of CO_2 emissions from the power industry. It will also result in an improved security of supply and efficient transmission of electricity. The project may also bridge the gap between supply and demand in developing countries.

The shortcomings of such a supply grid include difficulty in disconnecting and fixing an entire HVDC grid in the case of a single fault; difficulty in applying a cost–benefit methodology that efficiently allocates the investment cost of the grid between countries; the link between electricity markets and grid capacities needing to be revised; a fear of dependence on other countries for one's own electricity supply; and the public's resistance to extensions in the grid that have the potential of adversely impacting their health, property and environment.

3.6 Synergies and hindrance

A number of challenges are present in the development of an energy efficient system. These include the inability to develop a reliable and robust system that fulfils the demand of the market. The energy sector is in a constant state of flux, and advances in the realm of technology are creating increased demand for electricity. How technological advancements will shape the future demand for energy is uncertain. For instance, the rapid advancements in shale gas extraction could create great reverberations in the economy and the environment in general and energy sector in particular. A *World Energy Outlook* report published in 2011 stated that development in energy systems is a costly process. The report said that investments in energy development will likely reach about $35 trillion; that is a large sum in the context of the risk levels. Energy systems are volatile due to the political, economic and social influences that present a great danger to global energy system security.

Notes

1 A. Crosby (2006). *Children of the Sun: A History of Humanity's Insatiable Appetite for Energy.* New York: W. W. Norton.
2 W. C. Otto (2015). Future energy system development depends on the past learning opportunities. *Wires Energy and Environment,* 19 (10), 172.
3 B. Kroposki, M. O'Malley and S. Macmillan (2012). *Energy Systems Integration: A Convergence of Ideas.* National Renewable Energy Laboratory.
4 A. Crosby (2006). *Children of the Sun: A History of Humanity's Insatiable Appetite for Energy.* New York: W. W. Norton.
5 A. Grübler, N. Nakićenović and D. G. Victor (1999). Dynamics of energy technologies and global change. *Energy Policy,* 27 (5), 247–280.
6 U.S. Energy Information Agency (2015). *Annual Energy Outlook 2015 with Projections to 2040.* Accessed January 2016 from www.eia.gov/forecasts/aeo/pdf/0383(2015).pdf
7 B. Kovarik (1998). Henry Ford, Charles Kettering and the fuel of the future. *Automotive History Review,* 32, 7–27.
8 B. Kroposki, M. O'Malley and S. Macmillan (2012). *Energy Systems Integration: A Convergence of Ideas.* National Renewable Energy Laboratory.
9 M. Hogan (2013). Aligning power markets to deliver value. *The Electricity Journal,* 26 (8), 23–34.
10 CERA (2013). *Energy Vision 2013: Energy Transition Past and Future.* London: CERA.
11 IEA Statistics (2015). Accessed February 2016 from www.iea.org/publications/freepublications
12 J. C. Williams (2006). *History of Energy.* Accessed November 2015 from www.fi.edu.com
13 Vattenfall (2012). *Six Sources of Energy: One Energy System.* European Energy Association. Accessed January 2015 from https://corporate.vattenfall.com/globalassets/corporate/about_vattenfall/ generation/six_sources_of_energy_one_energy_system.pdf
14 C. McCain Nelson and A. Harder (2015). EPA emissions rule to mandate limits beyond proposed targets. *Wall Street Journal.* Accessed January 2015 from www.wsj.com/articles/epa-emissions-rule-to-mandate-limits-beyond-proposed-targets-1438488002
15 British Petroleum (2014). *BP Energy Outlook 2015.* Accessed December 2015 from www.bp.com/en/global/corporate/energy-economics/energy-outlook-2035.html
16 ASME (2009). *Vulcun Street Power Plant.* American Society of Mechanical Engineers.
17 S. Aggarwal and E. Burges (2014). Performance based models to address utility challenges. *The Electricity Journal,* 27 (6), 48–60.
18 International Energy Agency (IEA) (2012). *Energy Technology Perspectives.* International Energy Agency.
19 W. C. Otto (2015). Future energy system development depends on the past learning opportunities. *Wires Energy and Environment,* 19 (10), 172.
20 NETL (2009). *History of U.S. Coal Use.* Accessed November 2015 from www.netl.doe.gov
21 C. McCain Nelson and A. Harder (2015). EPA emissions rule to mandate limits beyond proposed targets. *Wall Street Journal.* Accessed February 2016 from www.wsj.com/articles/epa-emissions-rule-to-mandate-limits-beyond-proposed-targets-1438488002
22 IEA (2012). *Energy Technology Perspectives.* International Energy Agency.
23 U.S. Energy Information Agency (2015). *Annual Energy Outlook 2015 with Projections to 2040.* Accessed February 2016 from www.eia.gov/forecasts/aeo/pdf/0383(2015).pdf
24 Ibid.
25 ERP (2011). *The Future Role for Energy Storage in the UK: Executive Summary and Conclusions.* Accessed May 2016 from www.epri.com/abstracts/Pages/ProductAbstract.aspx?ProductId=000000003002000312
26 A. Zucker, T. Hinchliffe and A. Spisto (2013). *Assessing Storage Value in Energy Markets: A Literature Review.* JRC Scientific and Policy Reports, Report EUR 26056 EN.
27 P. G. Taylor, R. Bolton, D. Stone and P. Upham (2013). Developing pathways for energy storage in the UK using a co-evolutionary framework. *Energy Policy,* 63, 230–243.
28 C. Spataru, Y. C. Kok, M. Barrett and T. Sweetnam (2016). Techno-economic assessment for optimal energy storage mix. *Energy Procedia,* 83, 515–524.
29 P. H. Grünewald, T. T. Cockerill, M. Contestabile and P. J. Pearson (2012). The socio-technical transition of distributed electricity storage into future networks: System value and stakeholder views. *Energy Policy,* 50, 449–457.

30 G. Strbac, M. Aunedi, D. Pudjianto, P. Djapic, F. Teng, A. Sturt, D. Jackravut, R. Sansom, V. Yufit and N. Brandon (2012). *Strategic Assessment of the Role and Value of Energy Storage Systems in the UK Low Carbon Energy Future*. The Carbon Trust.

31 P. Grünewald, T. Cockerill, M. Contestabile and P. Pearson (2011). The role of large scale storage in a GB low carbon energy future: Issues and policy challenges. *Energy Policy*, 39 (9), 4807–4815.

32 D. Bhatnagar and V. Loose (2012). *Evaluating Utility Owned Electric Energy Storage Systems: A Perspective for State Electric Utility Regulators* (No. SAND2012-9422). Sandia National Laboratories.

33 P. Denholm, J. Jorgenson, M. Hummon, T. Jenkin, D. Palchak, B. Kirby and M. O'Malley (2013). The value of energy storage for grid applications. *Contract*, 303, 275–300. Accessed May 2014 from www.nrel.gov/docs/fy13osti/58465.pdf

34 R. Sioshansi, P. Denholm, T. Jenkin and J. Weiss (2009). Estimating the value of electricity storage in PJM: Arbitrage and some welfare effects. *Energy Economics*, 31 (2), 269–277.

35 R. Loulou, U. Remme, A. Kanudia, A. Lehtila and G. Goldstein (2002). *Documentation for the TIMES Model, Part I. Energy Technology Systems Analysis Programme (ETSAP)*. Accessed May 2016 from www.etsap.org/documentation.asp

36 C. Heaton (2014). *Modelling Low-Carbon Energy System Designs with the ETI ESME Model*. Accessed May 2016 from www.eti.co.uk/wp-content/uploads/2014/04/ESME_Modelling_Paper.pdf

37 P. Grünewald, T. Cockerill, M. Contestabile and P. Pearson (2011). The role of large scale storage in a GB low carbon energy future: Issues and policy challenges. *Energy Policy*, 39 (9), 4807–4815.

38 Ibid.

39 G. Strbac, M. Aunedi, D. Pudjianto, P. Djapic, F. Teng, A. Sturt, D. Kackravut, R. Sansom, V. Yufit and N. Brandon (2012). *Strategic Assessment of the Role and Value of Energy Storage Systems in the UK Low Carbon Energy Future*. Report for Carbon Trust, Energy Futures Lab, Imperial College, London. Accessed from https://workspace.imperial.ac.uk/energyfutureslab/Public/Strategic%20Assessment%20of%20the%20Role%20and%20Value%20of%20Energy%20Storage%20in%20the%20UK.pdf

40 DSIM Data Sheet (2013). Accessed May 2016 from www.acicomms.com/wordpress/wp-content/uploads/2013/12/08.-DSIM-Data-sheet-Rev-H-150504.pdf

41 EPRI (Electric Power Research Institute) (2013). *Cost-Effectiveness of Energy Storage in California: Application of the EPRI Energy Storage Valuation Tool to Inform the California Public Utility Commission*. Proceeding R. 10-12-007, 3002001162 Technical Update, June 2013.

42 R. Sioshansi, P. Denholm, T. Jenkin and J. Weiss (2009). Estimating the value of electricity storage in PJM: Arbitrage and some welfare effects. *Energy Economics*, 31 (2), 269–277.

43 British Petroleum (2014). *BP Energy Outlook 2015*. Accessed December 2015 from www.bp.com/content/dam/bp/pdf/Energy-economics/Energy-Outlook/Energy_Outlook_2035_booklet.pdf

44 R. Loulou, U. Remme, A. Kanudia, A. Lehtila and G. Goldstein (2002). *Documentation for the TIMES Model, Part I. Energy Technology Systems Analysis Programme (ETSAP)*. Accessed May 2016 from www.etsap.org/documentation.asp

45 P. Grünewald, T. Cockerill, M. Contestabile and P. Pearson (2011). The role of large scale storage in a GB low carbon energy future: Issues and policy challenges. *Energy Policy*, 39 (9), 4807–4815.

46 R. Sioshansi, P. Denholm, T. Jenkin and J. Weiss (2009). Estimating the value of electricity storage in PJM: Arbitrage and some welfare effects. *Energy Economics*, 31 (2), 269–277.

47 J. Ellison, D. Bhatnagar and B. Karlson (2012). *Maui Energy Storage Study*. SAND2012-10314. Albuquerque, NM: Sandia National Laboratories.

48 V. Viswanathan, X. Guo and F. Tuffner (2010). *Energy Storage for Power Systems Applications: A Regional Assessment for the Northwest Power Pool (NWPP)* (Vol. 19300). Pacific Northwest National Laboratory.

49 DSIM Data Sheet (2013). Accessed May 2016 from www.acicomms.com/wordpress/wp-content/uploads/2013/12/08.-DSIM-Data-sheet-Rev-H-150504.pdf

50 Low Carbon Innovation Coordination Group (LCICG) (2012). *Technology Innovation Needs Assessment (TINA) Electrcity Networks & Storage (EN&S) Summary Report, August 2012*. Accessed May 2014 from www.lowcarboninnovation.co.uk/document.php?o=15

51 P. Grünewald, T. Cockerill, M. Contestabile and P. Pearson (2011). The role of large scale storage in a GB low carbon energy future: Issues and policy challenges. *Energy Policy*, 39 (9), 4807–4815.

52 J. Eyer and G. Corey (2013). *Energy Storage for the Electricity Grid: Benefits and Market Potential Assessment Guide*. Sandia National Laboratories.

53 R. Sioshansi, P. Denholm, T. Jenkin and J. Weiss (2009). Estimating the value of electricity storage in PJM: Arbitrage and some welfare effects. *Energy Economics*, 31 (2), 269–277.

54 A. Nourai (2009). *Why and How Electric Vehicle Li-ion Batteries are Penetrating the Utility Market*. EV Li-ion Battery Forum, 2–3 September, Shanghai, China.

55 R. Walawalkar and J. Apt (2008). *Market Analysis of Emerging Electric Energy Storage Systems*. National Energy Technology Laboratory, 1–118.

56 R. Sioshansi, P. Denholm, T. Jenkin and J. Weiss (2009). Estimating the value of electricity storage in PJM: Arbitrage and some welfare effects. *Energy Economics*, 31 (2), 269–277.

57 E. Spahic, G. Balzer, B. Hellmich and W. Munch (2007, July). Wind energy storages – possibilities. In *Power Tech*, 2007 IEEE Lausanne (pp. 615–620).

58 G. N. Bathurst and G. Strbac (2003). Value of combining energy storage and wind in short-term energy and balancing markets. *Electric Power Systems Research*, 67, 1–8.

59 DTI (2004). Accessed February 2015 from www.gov.uk/government/uploads/system/uploads/attachment_data/file/272133/6536.pdf

60 J. Barton and D. Infield (2004). Energy storage and its use with intermittent renewable energy. *IEEE Transactions on Energy Conversion*, 19 (2), 441–448.

61 G. Strbac, A. Shakoor, M. Black, D. Pudjianto and T. Bopp (2007). Impact of wind generation on the operation and development of the UK electricity systems. *Electric Power Systems Research*, 77, 1214–1227.

62 P. Grünewald, T. Cockerill, M. Contestabile and P. Pearson (2011). The role of large scale storage in a GB low carbon energy future: Issues and policy challenges. *Energy Policy*, 39 (9), 4807–4815.

63 P. C. Del Granado, S. W. Wallace and Z. Pang (2014). The value of electricity storage in domestic homes: A smart grid perspective. *Energy Systems*, 5 (2), 211–232.

64 C. Bullough, C. Gatzen, C. Jakiel, M. Koller, A. Nowi and S. Zunft (2004). Advanced adiabatic compressed air energy storage for the integration of wind energy. In *Proceedings of the European Wind Energy Conference*, EWEC 2004, 22–25 November, London.

65 L. Chen (2009). *China's Petroleum Industry*. Accessed November 2015 from www.world-energysource.com

66 F. Diaz-Gonzalez, A. Sumper, O. Gomis-Bellmunt and R. Villafáfila-Robles (2012). A review of energy storage technologies for wind power applications. *Renewable and Sustainable Energy Reviews*, 16 (4), 2154–2171.

67 H. L. Ferreira, R. Garde, G. Fulli, W. Kling and J. P. Lopes (2013). Characterisation of electrical energy storage technologies. *Energy Journal*, pp. 288–298.

68 IMECHE (Institute of Mechanical Engineers) (2014). *Energy Storage: The Missing Link in the UK's Energy Commitments*. Accessed January 2015 from https://www.imeche.org/docs/default-source/1-oscar/reports-policy-statements-and-documents/energy-storage---the-missing-link-in-the-uk-39-s-energy-commitments.pdf?sfvrsn=0

69 H. Ibrahim, A. Ilinca and J. Perron (2008). Energy storage systems: Characteristics and comparisons. *Renewable and Sustainable Energy Reviews*, 12 (5), 1221–1250.

70 J. Leadbetter and L. G. Swan (2012). Selection of battery technology to support grid-integrated renewable electricity. *Journal of Power Sources*, 216, 376–386.

71 A. Sparacino, G. F. Reed, R. J. Kerestes and Z. T. Smith (2012). *A Survey of Battery Energy Storage Systems and Modeling Techniques*. Power and Energy Society General Meeting, 2012, IEEE.

72 S. Sundararagavan and E. Baker (2012). Evaluating energy storage technologies for wind power integration. *Solar Energy*, 86 (9), 2707–2717.

73 C. Spataru and P. Bouffaron (2016). Off-grid energy storage. In T. Letcher (ed.), *Storing Energy with Special Reference to Renewable Energy Sources* (pp. 477–497). Berlin: Elsevier.

74 C. Spataru, Y. C. Kok, M. Barrett and T. Sweetnam (2016). Techno-economic assessment for optimal energy storage mix. *Energy Procedia*, 83, 515–524.

75 Gobitec and Asian super grid for renewable energies in Northeast Asia (2014). Accessed January 2015 from www.energycharter.org/fileadmin/DocumentsMedia/Thematic/Gobitec_and_the_Asian_Supergrid_2014_en.pdf

4 Brief overview of energy systems models and methodologies

A large number of energy demand and supply methodologies have been developed over the course of time that help in analysing systems as well as subsystems. These models and tools are developed to gain deep insight into the current and future demand and supply interactions in the energy sector.

The energy systems models and methodologies give a better understanding of the interaction between the environment, economy and energy. This information can be used to make policies and plans to ensure that energy demands are fulfilled in the most cost-effective manner. In this chapter we will take a brief look at some effective models and methodologies that are used to analyse energy systems.

4.1 Current developments in energy systems models

Energy systems models are based on analytical and theoretical methods from different disciplines. These include engineering, operations, research, management, economics and science. The models make use of several techniques such as econometrics, mathematical programming, and other relevant methods relating to statistical analysis and network analysis.[1]

Soft linking models represent the next step towards a whole system approach that covers several sectors and subsectors and considers demand and supply constraints, geographical areas and regions.

A number of detailed reviews have been written on various aspects of the energy systems models, ranging from comparative to systematic review. A detailed list of multiple reviews is presented below that is not exhaustive.

1 Hoffman and Wood (1976)[2]
2 Wirl and Szirucsek (1990)[3]
3 Markandya (1990)[4]
4 Pandey (2002)[5]
5 Nakata (2004)[6]
6 Jebaraj and Iniyan (2006)[7]
7 Urban *et al.* (2007)[8]
8 Bahn *et al.* (2009)[9]
9 Bhattacharyya and Timilsina (2010)[10]
10 Herbst *et al.* (2012)[11]
11 Mancarella (2014)[12]
12 Hall and Buckley (2016).[13]

Reviews by Hoffman and Wood (1976) and Wirl and Szirucsek (1990) offer a deep insight into the developments and advancement in energy modelling. The former introduced the energy accounting approach that has been used in the modelling of energy systems in the US since the 1950s. Examples of models that are based on this approach include Modele d'Evolution de la Demande d'Energie (MEDEE) or Long-range Energy Alternative Planning (LEAP) and the Model for Analysis of Energy Demand. Both models capture supply chain activities that offer environmental, economic and resource analysis. An advantage of the approach is that it allows for the application of techniques to optimise access to different energy system frameworks that utilise alternative sources of energy and considers the demands of end users. Linear programming uses the same concept.

The Brookhaven Energy System Optimization Model (BESOM) that was introduced in 1977 is another effective model that is used to analyse the future allocation of resources. The model has been further developed to include an input–output table and a macro-economic linkage.[14] Most countries have adopted the BESOM model to develop their own technique for the analysis of energy systems.[15] These include ENERGETICOS in Mexico, the TERI Energy Economy Environment Simulation Evaluation model in India, and the Regional Energy Scenario Generator (RESGEN).

In 1974 Hudson and Jørgenson developed for the first time an energy model that linked an inter-industry energy model with a macroeconomic growth model that made use of demand and output.[16]

Most of the above models were limited in application to a national level. The first contribution to developing an energy system model at a global scale was made with the application of a World Dynamics model developed by Jay Forrester.[17] Although the model was limited in representing the energy system, it marked a new era in energy modelling.

Major advances were made between 1973 and 1985 that analysed the interdependence between the economy and energy. A study published in 1979 explained the relationship between energy and substitution using elasticity of substitution.[18] Another study published during the same year proposed that in the short run energy and capital are complementary, while in the long run they are substitutable.[19]

Energy models that were developed in the 1980s and 1990s addressed environmental concerns. The models addressed the energy–environment interaction, focusing on environmental costs and alternative scenarios. At the same time attention was also diverted towards global and regional models, which led to the development of such models as the RAINS-Asia model, Second Generation Model (SGM), Asian-Pacific Integrated Model (AIM), Global 2100, DICE and POLES.

The uncertainty present in the existing models, as well as the complexity due to the interrelationship between technology, behaviour and resources, led to the development of such initiatives as the VLEEM.

Markandya (1990) focused on planning models relating to energy systems to examine their ability to capture environmental concerns and also their appropriateness for developing countries. Nakata (2004) focused on environmental models, Pandey (2002) and Urban *et al.* (2007) compared different models, while Jebaraj and Iniyan (2006) offered a systematic comparison of different models and examined their ability to capture features of the energy market in the developing countries.

4.2 Grouping the underlying methodology of energy systems models

We can categorise energy systems models into different categories using multiple criteria. In this section we will look at differing approaches that are used to group the energy systems models.

The following are some of the approaches that have been suggested to categorise the underlying methodology of energy systems models:[20] input–output approaches; linear programming; econometric method; system dynamics; and game theory.

Table 4.1 shows a set of attributes that can be used to categorise energy models.[21]

The following are four additional approaches that have been suggested for the classification of models developed using the Meta-Net approach for energy system modelling and demand analysis:[22,23,24,25]

1   the modelling approach (top-down and bottom-up);
2   methodology (partial equilibrium, general equilibrium or hybrid);
3   modelling technology (optimisation, econometric or accounting);
4   the spatial dimension (national, regional and global).

Below are five other classification methods that have been developed after comparing and analysing the models through a spatial focus and sector-wise coverage.[26]

*   *Type I*: Bottom-up, optimisation-based models (RESGEN (Regional Energy Scenario Generator); EFOM (Energy Flow Optimisation Model); MARKAL model (Market Allocation model); TIMES (The Integrated MARKAL-EFOM System); MESAP (Modular Energy System Analysis and Planning));
*   *Type II*: Bottom-up, accounting models (LEAP (Long-range Energy Alternatives Planning model));
*   *Type III*: Top-down, econometrics models (DTI energy model);
*   *Type IV*: Hybrid models (NEMS (National Energy Modelling System); POLES (Prospective Outlook on Long-term Energy Systems); WEM (World Energy Model); SAGE (System for the Analysis of Global Energy Markets));
*   *Type V*: Electricity planning (WASP (Wien Automatic System Planning); EGEAS (Electricity Generation Expansion Analysis System)).

Table 4.1 Classification of energy–economy models

Paradigm	Space	Sector	Time	Examples
Top-down/simulation	Global; national	Macro-economy, energy	Long term	AIM, SGM2, I/O models
Bottom-up optimisation/ accounting	National, regional	Energy	Long term	MARKAL, LEAP
Bottom-up optimisation/ accounting	National, regional, local	Energy	Medium term, short term	Sector models (power, coal)

Apart from the above methods of model classification, studies have suggested various other methodologies that include the following:[27]

- econometric models
- macro-economic models
- economic equilibrium models
- resource allocation models
- optimisation models
- accounting models
- simulation models
- backcasting models
- multi-criteria models.

Metrics can also be used for the classification of energy systems models that cover different aspects. A model should be categorised based on multiple dimensions such as topology, model equations, spatial and temporal dimensions, parameter data, modelled system constraints, input/output information and model purpose.

Notes

1 K. Hoffman and D. O. Wood (1976). Energy system modelling and forecasting. *Annual Review of Energy*, 1, 423–453.
2 Ibid.
3 F. Wirl and E. Szirucsek (1990). Energy modelling – a survey of related topics. *OPEC Review*, Autumn, pp. 361–378.
4 A. Markandya (1990). Environmental costs and power system planning. *Utilities Policy*, October, pp. 13–27.
5 R. Pandey (2002). Energy policy modelling: Agenda for developing countries. *Energy Policy*, 30, 97–106.
6 T. Nakata (2004). Energy-economic models and the environment. *Progress in Energy and Combustion Science*, 30, 417–478.
7 S. Jebaraj and S. Iniyan (2006). A review of energy models. *Renewable and Sustainable Energy Reviews*, 10 (4), 281–311.
8 F. Urban, R. M. J. Benders and H. C. Moll (2007). Modelling energy systems for developing countries. *Energy Policy*, 35 (6), 3473–3482.
9 O. Bahn, A. Haurie and D. S. Zachary (2009). Mathematical modelling and simulation methods in energy systems, Vol. II. In *Encyclopedia of Life Support Systems (EOLSS)*. Edited by J. A. Filar and J. B. Krawczyk. Oxford: EOLSS Publishers Co Ltd.
10 S. C. Bhattacharyya and G. R. Timilsina (2010). *A Review of Energy Systems Models*. Accessed May 2016 from www.ewp.rpi.edu/hartford/~ernesto/S2013/MMEES/Papers/ENERGY/1EnergySystemsModeling/Bhattacharyya2010-ReviewEnergySystemModels.pdf
11 A. Herbst, F. Toro, F. Reitze and E. Jochem (2012). *Introduction to Energy Systems Modelling*. Accessed May 2016 from http://irees.eu/irees-wAssets/ docs/publications/journal-reviewed/Herbst-et-al-2012_Introduction-to-Energy-Systems-Modelling_SJES.pdf
12 P. Mancarella (2014). MES (multi-energy systems): An overview of concepts and evaluation models. *Energy*, 65, Feb, 1–17.
13 L. M. H. Hall and A. R. Buckley (2016). *A Review of Energy Systems Models in the UK: Prevalent Usage and Categorization*. Accessed May 2016 from www.sciencedirect.com/science/article/pii/S0306261916301672
14 K. Hoffman and D. O. Wood (1976). Energy system modelling and forecasting. *Annual Review of Energy*, 1, 423–453.
15 M. Munasinghe and P. Meier (1993). *Energy Policy Analysis and Modelling*. Cambridge: Cambridge University Press.

16 E. A. Hudson and D. W. Jørgenson (1974). US energy policy and economic growth: 1975–2000. *Bell Journal of Economics*, 5, 461–514.
17 J. Forrester. *Origin of System Dynamics*. Accessed January 2015 from www.systemdynamics. org/DL-IntroSysDyn/origin.htm
18 W. W. Hogan and A. S. Manne (1979). Energy–economy interactions: The fable of the elephant and the rabbit. *Advances in Economics of Energy and Resources*, 1, 7–26.
19 E. R. Berndt and D. O. Wood (1979). Engineering and econometric interpretations of energy–capital complementarity. *American Economic Review*, 69 (3), 342–352.
20 K. Hoffman and D. O. Wood (1976). Energy system modelling and forecasting. *Annual Review of Energy*, 1, 423–453.
21 R. Pandey (2002). Energy policy modelling: Agenda for developing countries. *Energy Policy*, 30, 97–106.
22 A. Markandya (1990). Environmental costs and power system planning. *Utilities Policy*, October, pp. 13–27.
23 M. Kanagawa and T. Nakata (2006). Analysis of the impact of electricity grid interconnection between Korea and Japan – feasibility study for energy network in Northeast Asia. *Energy Policy*, 34, 1015–1025.
24 S. Ashina and T. Nakata (2008). Quantitative analysis of energy-efficiency strategy on CO_2 emissions in the residential sector in Japan – case study of Iwate prefecture. *Applied Energy*, 85, 204–217.
25 H. Wang and T. Nakata (2009). Analysis of the market penetration of clean coal technologies and its impacts in China's electricity sector. *Energy Policy*, 37, 338–351.
26 S. C. Bhattacharyya and G. R. Timilsina (2010). A review of energy system models. *International Journal of Energy Sector Management*, 4 (4), 494–518.
27 L. M. H. Hall and A. R. Buckley (2016). A review of energy systems models in the UK: Prevalent usage and categorisation. *Applied Energy*, 169 (1), 607–628.

5 Discrepancies in energy systems models

Actual vs forecast

Energy demand and forecast have been extensively analysed in different research publications. A US-based study conducted in 2002 found that the forecasts about energy demands generally exceeded the demand by 100 per cent.[1]

Most of the models that are generally used cover analytical assumptions that are in the form of 'black boxes'. In other words, they are hard to examine and the outcomes are difficult to imitate. The view that the models will offer more accurate results when a broad set of data is included is proved not to be valid at all times. Sometimes simple models can offer more precise outcomes as compared to complicated techniques.[2]

A number of discrepancies are present when it comes to modelling actual data and results.[3] Some of these include inaccurate representation of economic agent behaviour, incomplete coverage of environmental and social impacts, lack of adequate technical details, and unrealistic assumptions, including full and efficient application of allocated resources.

Additionally, representation of energy models for developing countries is another problem relating to data modelling. The existing models are relevant for industrial nations,[4] but they have not been found to be true for developing countries. Developing and developed countries have different energy constraints and other characteristics. For instance, China has become a major economic power, and as a result the focus on energy development has shifted from developed to developing countries. In fast-growing developing countries such as China, access to affordable and clean energy remains a major issue.

Scenarios in the global energy sector are complex and assimilate different assumptions. Table 5.1 depicts studies that cover different time horizons and geographical areas that were undertaken by non-governmental and inter-governmental institutions.

The studies shown in Table 5.1 cover different models and were based on multiple assumptions to show an accurate forecast of global energy systems. However, these studies did not offer details on the approach taken to address the issue. The *Global Energy Assessment* reports used the MESSAGE–IMAGE approach that consists of an engineering optimisation model suitable for medium- and long-term forecasts of the energy system, development of scenarios, and energy policy analysis. The approach is the core part of the modelling framework of the Energy Programme at IIASA. The advantage of the model is that it allows flexible information to be sourced for evaluating challenges in the sector that has been implemented for the development of energy scenarios.

Table 5.1 Selected studies reviewed and compared

Global studies	Situations	Time horizon
Global Energy Assessment, by IIASA (2012)	Supply, efficiency, mix	2050
Energy Technology Perspectives, by IEA (2014)	6DS, 4DS, 2DS	2050
World Energy Outlook, by IEA (2013)	New policies scenario 450 scenario Current policies scenario	2040
New Lens Scenarios, by Shell (2013)	Mountains, oceans	2050
BP Energy Outlook, by BP (2014)	–	2035
The Outlook for Energy: A View to 2040, by ExxonMobil (2013)	–	2040

Energy Perspectives, by Statoil (2014)	Reference scenario
	2 alternatives:
	  Low carbon
	  Policy paralysis
		2040

Global Renewable Energy Market Outlook, by Bloomberg New Energy Finance (2013)	Traditional territory, new normal, barrier bursting	2030
Energy [R]evolution, by Greenpeace and EREC (2013)	Reference, energy revolution	2050
2050 Global Energy Scenarios, by WEC (2013)	Jazz, symphony	2050

The *Energy Technology Perspectives* study listed in Table 5.1 made use of multiple models and combined backcasting and forecasting analysis to analyse about 500 multiple technologies. The *World Energy Outlook* study examined the data using the World Energy Model (WEM), which attempted to depict energy market operations.[5]

EREC and Greenpeace made use of Modular Energy System Analysis and Planning Environment (MESAP) to find out supply side results, while the Planning Network (PlaNet) was used to stimulate and forecast energy supply and demands, costs, and impact on the environment. The former is a system analysis tool, while the latter is a type of linear network module.[6]

WEC makes use of the Global MARKAL Model (GMM) that is a type of partial equilibrium structure, having a cost-minimising objective function and using a bottom-up approach. The model shows the overall energy system that is supported by the energy technology analysis programme of the IEA.

Presently more than half of the population of the world reside in cities. Increase in population will drive energy demands that will rise with the growing urbanisation trends up until 2050.[7,8] Figure 5.1 shows the WEC study that considered different alternative scenarios for growth in population for their Symphony and Jazz scenarios. Deviation between the trajectories are small but increase over time. By 2050 it is projected that the world population will stand at about 9 billion. For the period between 2020 and 2050, the highest deviation is depicted by the ETP study, which is followed by the Greenpeace and EREC study.

GDP growth is another major metric that affects energy systems demand. Figure 5.2 shows GDP grouped into four decades. To establish comparability levels at growth rates in GDP, growth rate remains at modest levels with no significant fluctuations.[9] Projections of GDP levels do not reflect political, financial and other disruptions as they cannot be projected with accuracy. Moreover, greater emphasis is made on China

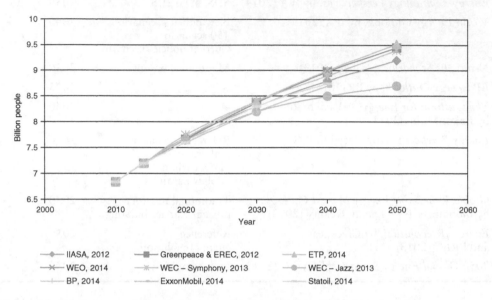

Figure 5.1 Comparison of population growth assumption in the reviewed studies.

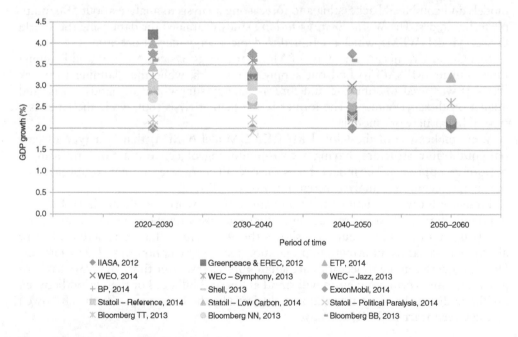

Figure 5.2 Comparison of GDP growth assumption in the reviewed studies.

and other Asian economies that do not follow trends that are similar to the developed Western economies.

ExxonMobil assumed the highest GDP growth forecasts until the year 2040, while Shell, Statoil and IIASA all assumed a scenario with modest economic growth and reduced improvements in the technologies.

The WEC Symphony scenario projects the highest economic growth as well as population growth between the years 2040 and 2050. IIASA and Greenpeace–EREC project the lowest level of economic growth.

5.1 Primary energy demand: a comparison

Primary energy demand refers to energy extracted from sources before it is converted, used and traded.[10] The capture of energy can occur without cleaning, separation or grading. The results for the demand of primary energy are shown in Figure 5.3.

A narrow distribution is projected for the year 2020. This is because targets and policies are already set with limited prediction of considerable political, social and economic assumptions. Projections vary from 685EJ for ExxonMobil to 530EJ for IIASA. The highest projection is made by Shell's Ocean scenario that assumes 777EJ for the same year. The lowest projection is made by the Energy Revolution study with 527EJ in the year 2030 that projects a world with energy-efficient appliances and environmentally aware consumers.

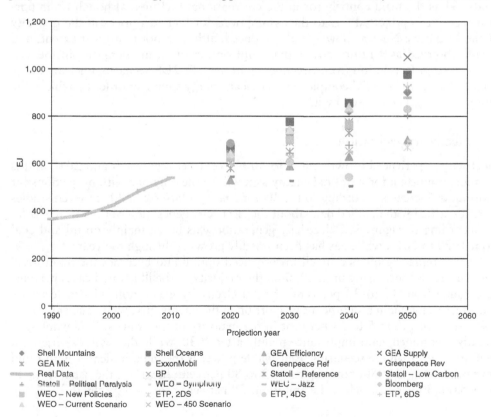

Figure 5.3 Comparison of primary energy demand projections in the reviewed studies.

IIASA's GEA Supply scenario projected the highest demand for 2040 and 2050. The projections will reach 1,050EJ in the year 2050, which is significantly different when compared to other studies.

5.2 Primary energy mix

The contribution of energy resources is presented in Figure 5.4. Although an attempt was made to gather comparable data, most studies did not offer information on the specific utility of hydropower, biomass and other mixes of fossil fuels. All scenarios project greater use of biomass and renewable energy after the year 2030.

The Statoil Low Carbon Scenario assumes 18 per cent of renewables penetration being the most 'optimistic scenario'. For 2030, 2040 and 2050, the Greenpeace Revolution Scenario includes the largest amount of renewables compared to the rest of the scenarios, projecting 24, 45 and 65 per cent respectively for every decade. On the other hand, Energy Technology Perspectives with the 6DS Scenario reflects the worst case scenario for the future world alongside 'Greenpeace Reference' and 'WEC Jazz' with less deployment of renewables. This is the case also for BP and ExxonMobil, which as oil industry companies promote perspectives with high levels of fossil fuels. Over the years a reduction in the use of oil and coal is observed, although they both remain active in the energy mix until 2050. Natural gas continues unabated as the main contributor in the conventional fuel mix, although no important increase is projected due to the assumptions for high gas prices in the majority of the scenarios. Nuclear power is also reduced, although not to a great extent, and is still observed as a future power (back-up) option of clean energy in all the scenarios, except Revolution from Greenpeace. In the ETP 2DS Scenario, nuclear energy is regarded as an essential complementary clean energy source in order to achieve the ambitious targets proposed by it.

5.3 Electricity generation

Electricity experienced growth equal to 40 per cent between 2000 and 2010 and is the most important form of final energy as it will be the only one with projections for continuous increases according to the *World Energy Outlook* in 2012. Seven studies were identified with detailed data about the electricity generation of the future world.

According to Figure 5.5, electricity generation was based mainly on oil and coal from 1970 to 2010, while gas has been steadily growing through the years.

Coal is replacing oil as a much lower-priced option and becomes the main fossil fuel. Gas is also increasing until 2050; in the majority of the illustrated cases percentages range from 15 to 30 per cent of total electricity generation. Hydroelectricity continues to maintain a significant proportion in the energy mix, while nuclear power possesses levels from 5 to 25 per cent in the majority of the studies. Renewables, as already mentioned, gain importance mainly after 2030, while they will experience a peak in 2050 in all the scenarios. Renewable power is the main indicator with rapid development until 2050, ranging from 12 to 80 per cent; however, the usual rates do not exceed 30 per cent below the ambitious 50 per cent proposed by the IEA.

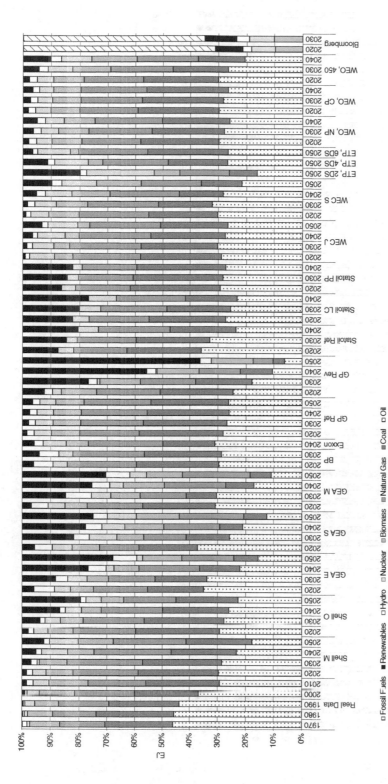

Figure 5.4 Comparison of primary energy mix scenario results and real data.

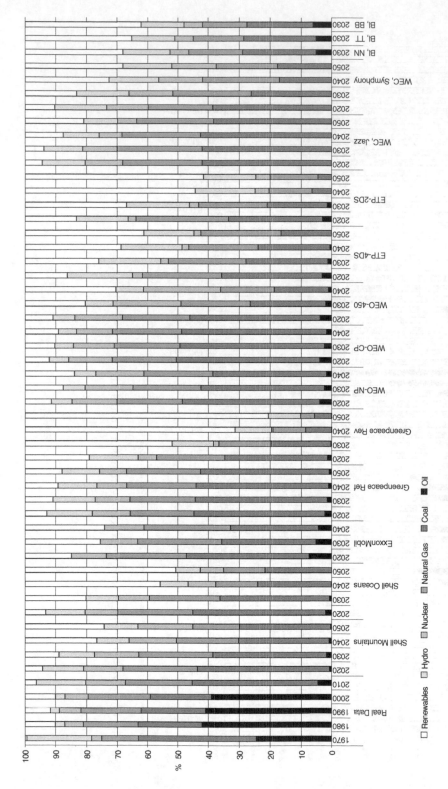

Figure 5.5 Comparison of electricity mix scenario results and real data.

5.4 Renewable energy share

Renewable energy growth was an original initiative of the European Union in 2007, with ambitious targets for 2020 and later for 2030 and 2050, but from 2020 the US and China will be the two major players in renewable energy investments.[11] In the future energy system, solar and wind energy (both onshore and offshore) are the resources that have the largest potential for expansion. Wind power increases with percentages ranging from 10 to 35 per cent depending on the scenario and is regarded as one of the most promising and secure investments in the energy sector. Photovoltaic installations, the main technology used for solar energy, can be found in smaller-scale projects such as residential or larger projects and have great potential for increase over the coming years, although they are more expensive. Solar power attains penetration of approximately 50–60 per cent in 2050 in scenarios such as Shell's 'Mountains' and the 'GEA Mix' pathway by IIASA (see Figure 5.6). Geothermal energy capacity varies in the scenarios and is observed mostly in the Greenpeace study; however, it is developed with a slower pace compared to the other resources, due to the limited availability of natural resources.

5.5 Environmental impact

The environmental impact of each scenario is presented in Figure 5.7 in the form of carbon dioxide emissions. All the assumptions and results mentioned above can be reflected in the amount of pollutant emissions, depending on each case's hypotheses. So far, CO_2 emissions have had a tendency to grow due to the extensive use of unsustainable energy; hence future scenarios appear to have been divided into two groups: those projecting significant environmental improvement and those with less optimistic projections for the environment.

In 2020, a projection year closer to the present time, values coincide, showing a relevant accuracy of the projections that are based on similar inputs. However, the Oceans Scenario from Shell and the 6DS Scenario from IEA show the highest projections (40 $GtCO_2$), while the Revolution Scenario from Greenpeace projects just 27 $GtCO_2$ (32.5 per cent less) as it is based mainly on efficient technologies, renewable energy and strong policies promoting sustainability.

Looking at 2030, the crowded results from 2020 are separating and the ETP 6DS Scenario is ranked first once more with the highest CO_2 emissions (44 $GtCO_2$), while the WEC Jazz Scenario (42 $GtCO_2$) follows. The most sustainable scenario is proposed by the Energy Revolution (20 $GtCO_2$). It is also observed that Shell's 'Mountains' does not show a wide deviation from 'Oceans', reflecting a quite similar approach until 2030, although these two scenarios are based on different principles. 'Oceans', which presents an unstable energy system, matches with WEC's 'Jazz' with high GDP assumptions and Statoil's 'Reference', and their assumptions lead to similar results close to 42–43 $GtCO_2$. The rest of the scenarios can be grouped into two categories close to 30 $GtCO_2$ and close to 40 $GtCO_2$, presenting a less sustainable energy system. In 2040, the picture is similar to 2030, but with relatively reduced carbon emissions close to 25 $GtCO_2$ and 40 $GtCO_2$. The WEO Current Scenario with less optimistic assumptions records 46 $GtCO_2$, while the 'Greenpeace Revolution' pathway is always the most environmentally friendly case.

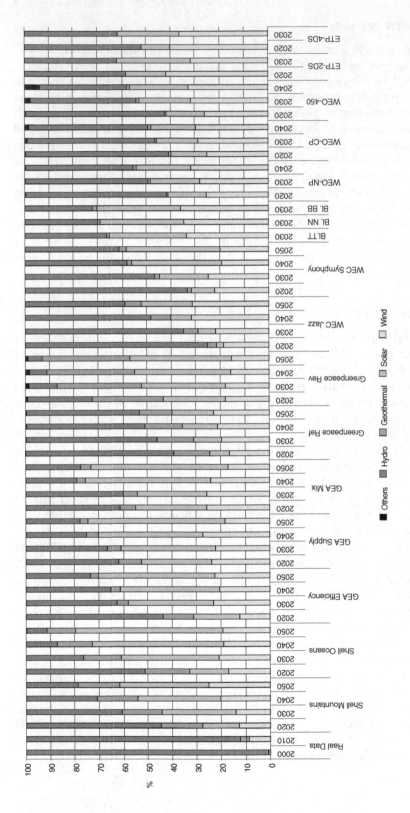

Figure 5.6 Comparison of renewable energy mix scenario results and real data.

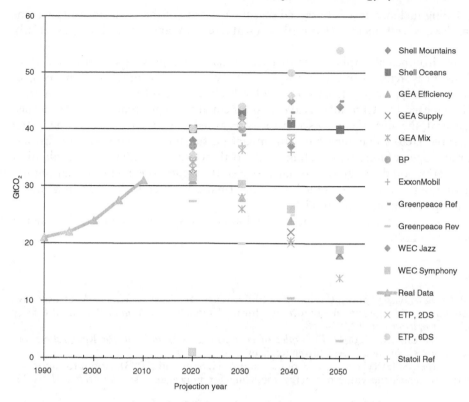

Figure 5.7 Comparison of CO$_2$ emissions scenario results.

In 2050, the results appear to be more dispersed, reflecting the number of uncertainties that are included in these kinds of long-term projections. In 2050, again '6DS' from IEA projects is considerably the largest amount of CO$_2$ emissions (54 GtCO$_2$). Yet again, WEC Symphony and GEA Efficiency and Supply have similar projections, while Greenpeace Reference continues to preserve its high scores in emissions, based only on low progress from the current policies. Finally, Energy Revolution presents an idealistic future world with CO$_2$ emission touching levels close to zero (3 GtCO$_2$). This is unrealistic, as already mentioned, as it requires a total transformation of the global economic and energy system.

5.6 Conclusions

The main objective of this exercise was to review and compare the latest released global energy scenario studies published between 2012 and 2014. Most of the scenarios have similar projections until 2020, and the main differentiations between them take place after 2030 or 2040. In addition, it was shown that the majority of the studies propose two or three different cases aiming to capture different perspectives, such as Shell, IIASA, Greenpeace, ETP, etc. The different views usually present a more environmentally friendly case and a reference scenario assuming no important diversifications from the present. On the other hand, studies mainly from industrial companies such as BP, ExxonMobil and Statoil suggest principally one future projection favouring the use of fossil fuels.

Regarding fuel mix, it was observed that the majority of the scenarios follow common pathways with increased use of renewable energy and natural gas, especially after 2030.

The environmental implications of each scenario are presented through carbon emissions measurements in the atmosphere. As we diverge from the present, projections vary greatly, ranging from very low levels close to zero for optimistic scenarios (e.g. Greenpeace-EREC) to increased emissions equal to 95 per cent from 2010 levels.

In general, the area of energy scenarios is regarded as very complicated due to the innumerable factors that may have an impact on the future energy system. In that sense, the scope of future work is huge and mostly related to model inputs evaluation and their effect on the scenario outputs. One of the limitations identified during the conducting of this exercise was the limited information related to simulation methods and assumption inputs in the models.

In conclusion, forecasting the future global energy system remains a challenging and uncertain task.

Notes

1 P. P. Craig, A. Gadgil and J. G. Koomey (2002). What can history teach us? A retrospective examination of long-term energy forecasts for the United States. *Annual Review of Energy and the Environment*, 27, 83–118.
2 J. S. Armstrong (ed.) (2001). *Principles of Forecasting: A Handbook for Researchers and Practitioners*. Norwell, MA: Kluwer Academic.
3 J. A. Laitner, S. J. DeCanio, J. G. Coomey and A. H. Sanstand (2003). Room for improvement: Increasing the value of energy modeling for policy analysis. *Utilities Policy*, 11, 87–94.
4 F. R. Urban, R. J. M. Benders and H. C. Moll (2007). Modelling energy systems for developing countries. *Energy Policy*, 35, 3473–3482.
5 T. Topalgökçeli and F. Keşicki (2013). *World Energy Model*. Paris: IEA.
6 D. Connolly, H. Lund, B. V. Mathiesen and M. Leahy (2010). A review of computer tools for analysing the integration of renewable energy into various energy systems. *Applied Energy*, 87, 1059–1082.
7 IIASA, Global Energy Assessment (GEA) (2012). *Toward a Sustainable Future*. Cambridge and New York: Cambridge University Press, and the International Institute for Applied Systems Analysis, Laxenburg, Austria.
8 ExxonMobil (2014). *The Outlook for Energy: A View to 2040*. Accessed April 2016 from http://cdn.exxonmobil.com/~/media/global/files/outlook-for-energy/2016/2016-outlook-for-energy.pdf
9 World Bank (2013). *GDP Growth (Annual %)*. Accessed May 2014 from http://data.worldbank.org/indicator/NY.GDP.MKTP.KD.ZG
10 S. Øvergaard (2008). *Definition of Primary and Secondary Energy in 13th Meeting of the London Group*. Brussels: EUROSTAT.
11 R. J. Campbell (2014). *China and the United States – A Comparison of Green Energy Programs and Policies*. Congressional Research Service: Informing the legislative debate since 1914. Accessed from https://fas.org/sgp/crs/row/R41748.pdf

6 Drivers and challenges for energy systems modelling for the twenty-first century and of future WES integration

A lot of obstacles need to be overcome to pave the way for reduced global carbon emissions in the next few years. This chapter will review some of the challenges that are present in the decarbonisation of the economy. Moreover, here you will also read about the security of the supply system to create pathways for low economic generation, transmission and the carbon economy.

6.1 Drivers of the future whole system scene

The following will offer insight into the energy system that can be implemented practically. But before delving into this topic, we first have to understand the evolution of the dynamics of multiple components and also the goals that need to be set in terms of availability, accessibility and acceptability.

Looking at the entire picture is important as it will help in understanding the critical links between different components of the energy mix that include gas, electricity, heat and transport. If consumers change from heating their homes with gas to making use of electric heating, this will have greater implications.

There are two main stages that need to be considered: long-listing drivers of change, and short-listing drivers of change. The long-listing drivers of change include environmental, social, economic, legal, political and ethical drivers. These drivers will shape the needs of the future energy system at domestic, European and global levels.

Short-listing drivers of change consist of only those drivers that have the greatest impact on the consumers and the regulations. Energy demand, GDP and supply are the three main drivers that play an important role in achieving a sustainable energy development goal. Previous trends show that accessibility of energy is one of the main components of economic development. The process was introduced during the initial years of the Industrial Revolution but has somehow slowed down in many countries in the past three decades.

Previous trends also show that acceptability of energy is a main part of energy demand. The demand has evolved more towards a cleaner and more complex usage of energy. Moreover, availability is important for the two drivers, as sustained supply of energy shocks as well as crises will serve as obstacles in the development of the economy that will force society to adjust to a more costly environment.[1] However, global energy constraints prevent the introduction of efficient and affordable sources of energy that are also safe for the environment.

6.2 The changing global energy scene

A lot of issues are present at a global level that necessitate creative solutions. Decades of economic growth have resulted in increased demand for energy. The demand for energy is especially great in countries that have experienced an above-average annual growth rate of 2–3 per cent. This has put increasing pressure on oil and gas prices and was one of the major factors that contributed to a three to four-fold increase in energy prices.

International pundits have started believing that oil prices will remain over $100 per barrel, but the projection is not on target. Reduced economic activities, especially in the developing countries, resulted in a rapid decline in the price of oil that at the moment is hovering around $30–40 per barrel. This price trend, however, is expected to be short term. The price is expected to pick up when the economic activities of China and other major oil consumers again move into high gear.

That being said, due to monetary constraints, less developed economies do not have the same priorities. They do not have the monetary resources to make the switch to cleaner and cheaper sources of energy. This increases the need for greater cooperation at a global level to ensure that all countries are on an equal footing when it comes to cost-effective, clean and sustainable energy sources.

Talking about energy supply issues, there are a number of detailed concerns that need to be addressed. Some of these include volatility in the distribution systems, greater import of fossil resources, national control or monopoly over supply, access to affordable energy supplies, and renewable energy sources.

The present low-level fuel prices have provided relief to the economies of the less developed world. However, it has also reduced their urgency to switch their energy systems to one that results in lower levels of carbon emissions.

Figures 6.1 and 6.2 show electricity access in regions of the world and populations without access to electricity.

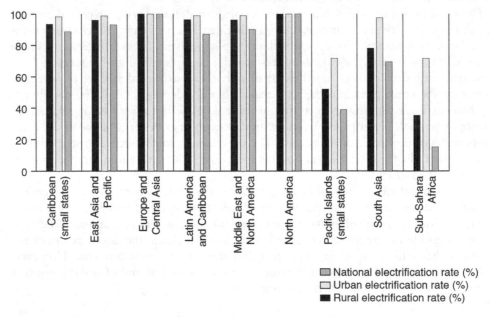

Figure 6.1 Electricity access in regions of the world in 2012. Data source: 2

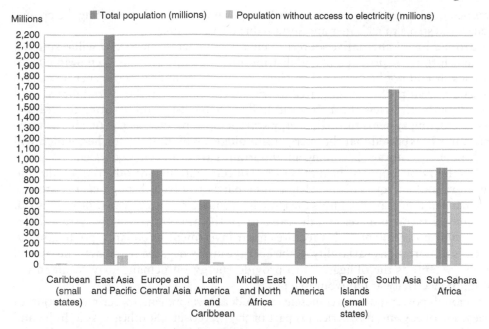

Figure 6.2 Total population and population without access to electricity. Data source: 3

Securing cleaner and stable sources of energy is a goal that needs to be shared by the developed, developing and less developed countries. The reason is that prolonged reliance on inefficient and costly energy sources will create economic disparities that will not be conducive for global trade. Stability of energy should be a major concern to ensure that all countries have equal opportunity to exploit technological advances that will lead to low product cost, greater income levels, more demand and therefore increased global trade, creating a win–win situation for all countries.

Talking about energy supply issues, there are a number of detailed concerns that need to be addressed. Some of these include volatility in the distribution systems, greater import of fossil resources, national control or monopoly over supply, access to affordable energy supplies, and renewable energy sources.

Oil supply disruptions have traditionally been the major concern of policymakers. However, today there is a greater need to ensure the supply of a secure and stable source of energy that does less harm to the economy as well as the environment. The time is ripe to rid our reliance on the 'black gold' that has created chaos in the world, resulting in unprecedented wealth transfer on a massive scale to the oil-rich countries, as well as contributing to climate change that could bring our mother nature to her knees.

6.3 Challenges in the energy systems

A lot of international studies, including IEA's WEO and IPCC, have shown that policy and technical options are present that can help address the issue of carbon emissions. Apart from that, research and development findings are also known that can contribute to the transition to a cleaner source of energy.

That being said, very little attention has been given with regards to making the change in energy systems to a cleaner and sustainable energy source.

Introduction of alternative sources of energy to generate electricity will result in a marked shift in the production and distribution of energy systems. Renewable technologies should not be discussed en bloc. For instance, hydroelectric energy sources have a different profile as compared to wind technology. Level of variability should be dealt with effectively through ensuring flexible back-up and boosting the robustness of existing systems through energy storage options such as gas turbines. In the long run it is expected that the energy technologies will improve further, which will result in energy storage becoming an important part of the energy system.

Variability of supply is the number one concern with the introduction of alternative sources of energy. This problem can be partly resolved through energy storage systems. Another way to address the issue is to ensure resource availability. This can be achieved through increasing the capacity of the alternative energy system, especially wind power plants. This requires great investment to enhance predictability, which is important as variability of the energy supply negatively impacts the overall energy system, as the obligations to provide energy on demand may result in the selling of electricity at unfavourable prices.

Regional power pools are present or under development in certain regions of Africa, Europe, Central America (as part of the Mercosur), Southeast Asia, India and in the US at the state level. The long-term goal of achieving low carbon emissions or even carbon neutrality by the year 2050 will require advancement of renewable and clean sources of energy. There is a strong need for R&D at a national level today to come up with advanced energy storage systems, cost-effective transmission lines and clean energy sources that will pave the way for the introduction of greener and more sustainable energy system technologies in the future.

6.4 Concluding remarks

Over the past few years, the global economy has faced a large number of obstacles and challenges. The free market economy and globalisation have resulted in unprecedented global economic growth. However, financial crises in the last decade have changed the landscape and led to a growing concern for building a sustainable and robust economic system.

In a way this has also resulted in an increased focus on making the switch to a clean, secure and more cost-effective energy system. This is true for most of the industrialised and rapidly developing countries. However, the less developed countries are still facing basic constraints that make it difficult to make progress in the development of alternative energy sources. A number of viable options are present to address the issue of climate change and ensure energy security. However, this will require gradual progress in the overall design of the process.

Flexible and intelligent energy systems can help solve a lot of issues concerning energy system development. They can facilitate efficient usage of alternative renewable energy systems that cause minimum harm to the environment. Such systems are not just optional but are necessary to successfully make the transition to clean and secure energy sources that lead to a significant reduction in carbon emissions as well as providing energy supply security in different regions around the world. And this requires making short-term energy system policies combined with long-term research on the advancement of new energy sources, as well as a greater integration between different interaction parts.

Notes

1 WEC (2004). *Drivers of the Energy Scene*. Accessed May 2016 from www.worldenergy.org/wp-content/uploads/2013/01/PUB_Drivers_of_the_Energy_Scene_2004_WEC.pdf
2 World Bank.
3 World Bank.

7 Worldwide conversation about energy systems integration

A lot of potential is present in the integration of energy systems. Different projects that are underway in the development of energy systems around the world are the result of worldwide collaboration and conversation about addressing the issue.

Whole energy systems integration is the buzzword that holds the secret to solving issues relating to the advancement in clean energy systems. There has been an increased interest from different organisations including universities in gaining better understanding and exploring future renewable energy sources, as well as the challenges and outcomes of their implementation.

The potential application of the worldwide integration of energy systems is wide and is beyond the scope of this book. In this chapter we will look at some of the worldwide conversations about energy systems integration that will greatly aid in reducing the carbon footprint at a global level.

7.1 Worldwide conversation apropos energy systems integration

In Table 7.1 a list of a number of institutions and their research is provided. The list is not exhaustive. There are other institutes worldwide, and I encourage the reader to contact the author regarding other institutes worldwide that are dealing with research on energy systems integration.

The concept of ESI is applicable in different scales from residential to commercial and across multiple energy as well as non-energy dimensions. For instance, the electricity and water nexus, electricity and gas nexus, gas and transport nexus, and others. This is relevant from a technical, market and regulatory perspective.

7.2 Shortcomings of existing work

There are a number of issues of great importance that are not dealt with in detail in the current research work. For instance, the roles and responsibilities of groups and institutions such as the UN (ICAO), the EC and the IPCC are not explored in detail.

Moreover, the effects of international laws and treaties such as GATT and EC law are important areas that have not been delved into in detail in the worldwide conversation. Lastly, the issue of competition and taxation are not studied in detail as to how they act as a constraint to the advancement of energy-efficient systems. The following areas need attention:

1 general understanding at a scientific level of the impact of aircraft emissions on the atmosphere;
2 scope for changes at an operational level to decrease impacts on the environment;
3 reduction of the potential impact of technological advancement and air transport;
4 a detailed comparison of alternative long-distance modes to replace current air transport routes;
5 demand management potential;
6 the impact of changes in air transport patterns, technologies, taxes, etc. on the cost effectiveness of air transport;
7 the impact of liberalisation and deregulation on energy system development;
8 the logistics of the worldwide air transport system.

All of the above-mentioned areas need to be looked at to ensure the introduction of efficient energy systems that meet the demands of the masses and leave a minimum footprint on the environment.

Table 7.1 Table with selected worldwide conversation

Institution	Scope of research
National Renewable Energy Laboratory (NREL), US	Conducts research on sustainable and renewable energy.
UK National Renewable Energy Centre (NaREC)	Carries out research on various types of renewable energy.
UCL Energy Institute, UK	Energy systems integration; energy space and time; coupled energy infrastructures.
Skoltech Centre for Energy Systems, Russia	Four main thrusts: • smart and resilient grids • energy markets and regulation • coupled energy infrastructures (power, gas, heat) • power electronics and devices.
The Institute for Integrated Energy Systems (IESVIc) at the University of Victoria, BC, Canada	Conducts research on techno-economic simulation and optimisation of energy systems.
The School of Renewable Energy Science (RES), Iceland	Affiliated with the University of Iceland. Carries out research on different forms of renewable energy.
Norwegian Centre for Renewable Energy (SFFE), Norway	Carries out research on alternative sources of renewable energy.
Centre for Alternative Technology (CAT), UK	The institution conducts research on alternative sources of renewable energy.
Florida Solar Energy Center (FSEC), US	Conducts research on use of solar energy.
Plataforma Solar De Almeria (PSA), Spain	Located in Spain, the research facility conducts research on solar energy.
Wind Science and Engineering Research Center, US	Carries out research on wind energy.
US–China Clean Energy Research Center (CERC)	Conducts research on the different forms of clean energy alternatives.

Part II
Modelling
Challenges and discussions

Prologue

The energy system is currently undergoing considerable changes and slowly transitioning towards the development of new sources that are efficient and reliable. Part II of the book will explore in depth various examples across residential and continental scales and levels, as well as the many domains of energy modelling.

8 Why do mathematical modelling?

Although models are abstract representations of reality based on a particular concept or ideal, they enable us to gain a deep understanding of real-life entities and variables through simplified renditions of the natural world. Expressed as computer codes, formulas and equations, models allow us to understand the critical elements of a reality, while leaving out the unnecessary aspects. Let us look in more detail at how models are developed and how they shape our understanding of reality.

In social and non-social sciences, experts often use mathematical models to explain a particular reality. These are based on a number of assumptions about the following:

- variables (aspects that change according to a particular event);
- parameters (aspects that do not change and thus are independent of a particular event);
- functional forms (the nature of the relationship between variables and parameters).

When models are being constructed, the objective is to develop the idea and represent it first as a conceptual model and then as a quantitative model. The purpose of the conceptual model is to help us understand how the system operates, and is expressed as a model diagram, consisting of state variables and material flows or causal effects.

The conceptual model is then transformed into a quantitative model through the addition of equations. This results in a series of dynamic questions pertaining to each variable, which can either be studied in a detailed manner or converted into computer codes to seek numerical solutions to trajectories associated with state variables.

Doing so will allow the user to resolve many of the underlying assumptions through model analysis. Furthermore, it allows the user to know what essential data needs to be gathered through conducting simulation predictions as well as the various control strategies that can be put into effect.

However, since modelling is not an accurate depiction of reality, there appears to be a trade-off between specificity and realism. The more precise the model is, the more chances there are for it to be theoretical and not reflective of reality. This is why it is important that the modeller's goals conform to the objectives of the study and consider the challenges associated with data precision.

The same is true for us in our daily lives; when we choose to represent specific realities of the world at the expense of others, we are essentially making a model to understand reality. This reality is driven by intuition, experience and knowledge, and is often used to challenge other theories, models and concepts to stand the test of comprehensiveness.

Based on the sophistication of the integration of the whole energy system, we can begin to realise performance and dependability and achieve significant cost efficiencies and limit environmental impacts.

In the following, I will shed light on the current perspective of modelling, the possible future challenges that we will face, and a detailed account of many modelling examples.

9 Indicators and their role in models

Modelling indicators are derived from basic data. In the case of energy efficiency, for instance, data can pertain to consumption of energy, output of energy and system features.

The modeller can draw multiple indicators, each of which is suited for a particular purpose, ranging from simple to comprehensive. The type of indicator chosen will thus hinge on the objectives of the modeller and the requirements of the model.

The use of indicators and metrics is very important in energy policy reports, showcasing the detailed characteristics of an economy's demand and supply patterns compared to those of other countries. This is not only helpful in gaining an insight into the future developments of energy systems, but also proves to be instrumental in establishing benchmarks for the purpose of gauging the government's public sector management performance and influencing policymaking.

For instance, a detailed analysis of the various indicators on healthcare, education and social security will offer transparency into how the government utilises taxpayers' money for maximising public benefit.

What indicators you choose will ultimately depend on what you want to understand. Some of these are as follows:

- descriptive: energy/CO_2 shares, etc.;
- normalised: energy/GDP or per capita;
- structural: measures of activity, output:
    - financial or physical
    - production and consumption;
- time series and comparative;
- consequential: emissions (including CO_2);
- decomposition over time, among countries;
- causal indicators: driving factors:
    - price and income relations
    - demographic, geographical and saturation factors;
- approach solid and rewarding, but data and time demanding.

Some of the indicators listed above are discussed in the following sections.

9.1 Energy efficiency indicators

Current indicators of energy efficiency are thorough in terms of knowledge, tools and methodologies. You will find a range of initiatives to gauge energy efficiency projects across multiple sectors and countries conducted by a number of reputable international institutes and universities. Some of these include:

- IEA Energy Indicators project;
- WEC-ADEME Energy Efficiency Indicators and Policy Project;
- DOE/EIA (USA);
- World Bank (energy efficiency country reports on Vietnam, Turkey and Russia).

IEA's approach on energy emission indicators is taken from the conceptual indicators structure resembling a pyramid.[1] In this, multiple disaggregated data are used to make adjustments for climate changes driven by the following:

- activity levels (these include population, value added, house area, person-kilometre and tonne-kilometre);
- structure (industry mix, appliance ownership, combination of transport modes);
- intensity (energy use per activity);
- CO_2 emissions (resulting from changes in power sector efficiency, fuel mix and end-use fuel mix).

In my understanding, the IEA indicator initiative along with the ODYSSEE Europe database provides the most robust database of indicators corresponding to the drivers of energy trends over time between different countries.[2]

The IEA report 'Energy Efficiency Indicators: Essentials for Policy Making' can be accessed online.[3]

The hierarchy of the pyramid indicator conceptual structure for energy efficiency begins from the most detailed aspects (e.g. process efficiencies) where more data is required, to the least detailed at the very top (e.g. sectoral intensities). The level of detail depends on the data and information available.

The historical trend analysis technique is used by both WEC and IEA to evaluate the impact of past energy efficiency developments and compare it across various countries.

Out of the many approaches I observed, I found that the uniformity of energy efficiency indicators and their efficient utilisation in policy design are the most preferred tool for cross-country evaluations. For this reason, indicators need to be constructed to ensure there is an alignment between the policy question and the aim of the research.

Phylipsen (2010) argued for the importance of a roadmap that could pave the way for the many levels of aggregation as well as the sectors that the indicators corresponding to the energy drivers can determine.[4] This will allow the country to design the best policy instrument and choose the appropriate indicators.

The roadmap would serve as an important two-directional decision-making instrument. Its starting point can be an objective that needs to be accomplished or an area that needs to be addressed or understood. However, it is worth pointing out that, for many countries worldwide, only aggregate data is available as key information sources. The two-directional roadmap can be accessed online.[5]

9.2 Resource efficiency indicators

One of the first examples of resource efficiency indicators is the initiative spearheaded by the European Union for a resource-efficient Europe as part of the Europe 2020 strategy. In this Resource Efficiency Roadmap, a detailed action plan is laid out exploring the details of the future action's design and implementation.[6] The Resource Efficiency Roadmap's Annex 6 consists of a provisional set of indicators that are set out in a three-layer format. These are as follows:

- a headline indicator, which is the first layer that focuses on resource productivity and is calculated by dividing GDP by the domestic material consumption amount;
- a dashboard complementary macro indicator, which is the second layer that has a particular emphasis on measuring the impact of resources and the environment;
- a group of context-specific indicators to evaluate the progress of specific goals and objectives.

A second crucial initiative was a joint project conducted by UNEP (United Nations Environment Programme), APRSCP (Asia-Pacific Roundtable on Sustainable Consumption and Production), CSIRO (Commonwealth Scientific and Industrial Research Organisation) and the EU's SWITCH-Asia Programme. As a result of their joint efforts, a thorough database is now available that features more than 20 countries in the Asia-Pacific region along with four decades' worth of resources at: http://uneplive. org. The detailed database comprises 118 indicators used to access nature resource reserves in the last 40 years across 26 countries in the Asia-Pacific region.

The report titled 'Indicators for a Resource Efficient Green Asia and the Pacific' can be accessed at: www.unep.org/roap/Activities/ResourceEfficiency/IndicatorsforaResource Efficient/tabid/1060186/Default.aspx

9.3 Indicators for sustainable development

The term 'sustainable development' was first defined in the Brundtland Report in 1987 as development that fulfils the strategic needs of the present without compromising those of future generations. Since then, a substantial amount of initiatives, campaigns and discussions have taken place to ensure there are sufficient resources to meet the needs of future generations. This also involves developing methodologies to track the progress of such efforts so as to ensure investments and programmes are being utilised efficiently and do not go off-track.

Overall, indicators are sources of information that can be either quantitative or qualitative; they play a pivotal role in ensuring all resources are being used effectively for developing a sustainable society.[7] The Human Development Index (HDI), for example, is one of the most reliable and accurate indicators for giving a detailed account of a person's standard of living and overall well-being.

There are also a range of indicators identified by Iddrisu and Bhattacharyya, in addition to the variety of sustainability indicator efforts outlined by government and non-government agencies.[8,9] For instance, the Compendium of Sustainable Development Indicator Initiatives, for sustainable development performance, identified nearly 600 indicators. Out of this number, 70 are international, 100 are at the national level, 70 are at the provincial or state level, while the rest lie at the local or metropolitan level.

Since then, a total of 17 sustainable development goals and nearly 170 indicators were agreed as of September 2015 and currently serve as the widely agreed framework for evaluating sustainable development.

The UN also approved the Inter-Agency and Expert Group's 231 indicators in March 2016 (IAEG-SDGs), a third of which are measurable. Figure 9.1 shows sustainable indicators from the local to global level.

a) *SDG indicators*

The IAEG-SDGs were established by the UN Statistical Commission in March 2015 for the purpose of developing the framework for sustainable development goals and monitoring them on a global scale. At the last meeting after the 47th Session of the UN Statistical Commission (UNSC 47), all of the approved 231 indicators were categorised into three separate tiers.

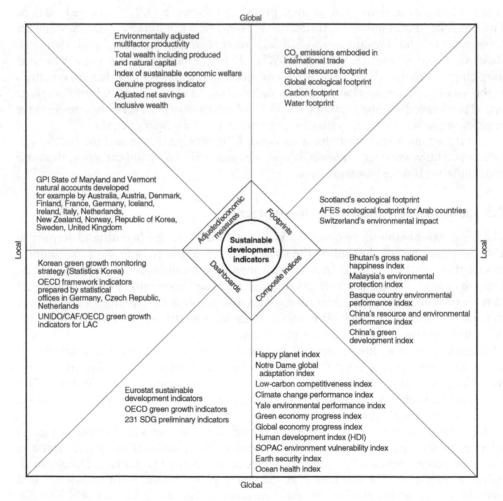

Figure 9.1 Sustainable development indicators.

The first tier includes a set of indicators for which a considerable amount of data and a structured methodology is available.

Tiers II and III, on the other hand, are those that raise interesting questions for the purpose of future policy decision making. Tier II, in particular, contains a list of indicators for which data is not widely available despite having a structured methodology; while Tier III contains no proper data nor does it contain an internationally approved methodology.

The objective of separating indicators into separate tiers is to ensure that both quantitative and qualitative factors are taken into consideration for a robust and comprehensive decision pertaining to sustainable development. The SDG index will thus be more of a dashboard through which countries can understand and evaluate their level with respect to the index and know the steps they should take for transforming their societies accordingly. It will provide countries with a quick snapshot in terms of how they rank in comparison with other countries that will provide them with the incentive to intensify their efforts towards achieving sustainable economies.

b) Measurement approaches

Efficiently measuring how well efforts are conforming to sustainable development goals requires a detailed measurement across a range of sectors and themes using the right combination of a set of indicators. The indicators used should be a mix of environmental, social and economic indicators in accordance with the particular needs outlined in the framework, one of which is that of the OECD.

From the comprehensive review of sustainability indicators outlined by Singh *et al.*, the GGKP in 2016 conducted a detailed analysis of the four factors that can be used together to measure the three dimensions of sustainability.[10] These four factors include dashboards of indicators, composite indices, environmental footprints and adjusted economic pressures.

At a more practical level, however, the methodology used to detail across sectorial monitoring of the goals and targets of sustainability is currently being developed by IAEG-SDGs.[11] The real challenge in sustainable development lies in the inter-linkages of various targets, resulting in a series of complex relationships that are impossible to pinpoint and dissect without using a large number of indicators or a set of complex composite indicators.

Stack Exchange Area 51[12] provides Open Modelling tools for SDGs showing how to use open source tools such as GIS, OnSSET, OSeMOSYS and CLEW.

9.4 Indicators for the resource nexus

Indicators for the resource nexus are used a lot differently than those for resource efficiency or for the development of sustainable development goals. Resource nexus indicators are utilised to understand the three following aspects:

1   use of resources (demand of resources by sector, efficiency and dependency);
2   nexus security (to determine which resources are scarce and what their location is, and also to measure the impact of sector–resource inter-linkages on securities);
3   opportunities between countries.

The nexus assessment can be conducted efficiently through the use of indicators in two important ways:

1   substantiating the analysis;
2   quantifying the specific issues and benefits of the outline by the input of stakeholders.

When assessing the findings, a special emphasis is given to inter-linkages between sectors (nexus) and the available resources, in addition to the relations between countries. This is to ensure that all countries are given an equal chance of utilising energy to meet their needs. For this, the cross-country resource nexus study starts with an understanding of the socio-economic profiles of particular countries and the region, as well as the geographical details and availability of resources.

The following is a list of indicators that are considered to be most important in conducting resource nexus studies:

1   World Development Indicators according to country as outlined in the World Bank Database, as well as other factors such as environment, society, demography, MDGs, economy, markets and states;
2   FAO-stat (Aquastat) water resources, agricultural management and usage of water;
3   UNECE Second Assessment of Trans-boundary Rivers, Lakes and Groundwaters according to water quality and water basin users;
4   Water Risk Atlas in terms of water basin users as provided by the Water Resources Institute Database and other essential variables including flood occurrence, seasonal variability, inter-annual variability, water stress and drought severity.

9.5 Spatial indicators (GIS)

In addition to indicators for resource nexus studies, spatial indicators also are useful for highlighting geographic information and hotspots. This consists of a number of activities performed at various levels with the use of databases. Here I will mention some of these pertaining to resource nexus aspects, such as energy, water and land:

*   nexus FAO database – according to country and/or by basin inter-linkages across water, energy and food sectors;
*   population density (NASA SEDAC) and boundaries (Global Administrative Areas);
*   global lakes and wetlands database (European Environment Agency);
*   industrial activities location (hydropower, power plants and heavy industry) (GRID ARENDAL);
*   urban areas (Global Rural–Urban Mapping Project, GRUMP – NASA/Columbia University);
*   deforestation/reforestation (global forest change as outlined by Hansen *et al.*[13]);
*   land use and land cover data (FAO's and JRC's databases), ecosystem-related activities;
*   protected areas and ecosystems (World Database on Protected Areas);
*   access to water and upstream protected land (FAO, AQUEDUCT global maps).

Notes

1 *IEA Energy Emission Indicators*. Accessed May 2016 from www.iea.org/publications/freepublications/publication/IEA_EnergyEfficiencyIndicators_EssentialsforPolicyMaking.pdf

2 *Energy Indicators Pyramid Data (2010)*. Accessed May 2016 from www.esmap.org/sites/esmap.org/files/EECI%20Energy%20efficiency%20indicators%20in%20developing%20country%20policy%20making%20July%202010.pdf

3 www.iea.org/publications/freepublications/publication/IEA_Energy EfficiencyIndicators_EssentialsforPolicyMaking.pdf

4 G. J. M. Phylipsen (2010). *Energy Efficiency Indicators: Best Practice and Potential Use in Developing Country Policy Making*. Accessed January 2015 from www.esmap.org/sites/esmap.org/files/EECI%20Energy%20efficiency%20indicators%20in%20developing%20country%20policy%20making%20July%202010.pdf

5 www.esmap.org/sites/esmap.org/files/EECI%20Energy%20efficiency%20indicators%20in%20developing%20country%20policy%20making%20July%202010.pdf

6 *Resource Efficiency*. Accessed May 2016 from http://ec.europa.eu/environment/resource_efficiency/about/roadmap/index_en.htm

7 R. Ciegis, J. Romanauskiene and G. Startiene (2009). Theoretical reasoning of the use of indicators and indices for sustainable development assessment. *Engineering Economics*, 3, 33–40.

8 I. Iddrisu and S. Bhattacharyya (2015). Sustainable Energy Development Index: A multidimensional indicator for measuring sustainable energy development. *Renewable and Sustainable Energy Reviews*, 50, 513–530.

9 T. M. Parris and R. W. Kates (2003). Characterising and measuring sustainable development. *Annual Review of Environment and Resources*, 28, 559–586.

10 R. K. Singh, H. R. Murty, S. K. Gupta and A. K. Dikshit (2012). An overview of sustainability assessment methodologies. *Ecological Indicators*, 15, 281–299.

11 *Indicators for the SDGs: Identifying Inter-linkages*. Accessed May 2016 from http://unsdsn.org/wp-content/uploads/2015/09/150816-Identifying-inter-linkages-SDSN-Briefing-for-IAEG.pdf

12 Accessed January 2016 from http://area51.stackexchange.com/proposals/97641/open-modelling-tools-for-sustainable-development-goals-sdgs?referrer=7NvWuKsreQkHJZVp15G0iQ2

13 M. C. Hansen, P. V. Potapov, R. Moore, M. Hancher, S. A. Turubanova, A. Tyukavina, D. Thau, S. V. Stehman, S. J. Goetz, T. R. Loveland, A. Kommareddy, A. Egorov, L. Chini, C. O. Justice and J. R. G. Townshend (2013). High-resolution global maps of 21st-century forest cover change. *Science*, 342 (6160), 850–853. Accessed November 2014 from https://earthenginepartners.appspot.com/science-2013-global-forest

10 Research challenges

Calculating the demand for building energy is a major challenge. This is because any kind of estimate requires measuring energy demands on an hourly basis that should cover a sufficient number of sample days to be representative for an entire year. To arrive at such a value requires a direct measurement or the use of a parameterised model.

10.1 Energy demand

Measuring directly is time-consuming, and using a parameterised model is impractical, owing to the introduction of new building systems and features such as large insulated windows. Furthermore, peak building energy demands are largely influenced by weather conditions and social activities, along with the performance of weather-dependent technology, such as solar collectors and heat pumps.

This is highly critical, as perhaps the biggest driver for the electricity network's total capacity is peak demand. Being unable to obtain a near-enough accurate measure will make it extremely difficult for organisations worldwide to achieve their resource nexus objectives.

To make things more challenging, the rise in peak demand will have an upward impact on the consumption of space heating electrification and of electric vehicles, owing to the diminishing reliance on fossil fuels as a source of renewable energy. This will pose a much larger obstacle in the effort to achieve harmony across energy, land and water.

Until then, a much more short-term plan is required to manage the burgeoning unpredictable demand. This could be the use of an active demand side management or response mechanism, which will seek to smooth out the gaps between demand and supply.

Storage systems can ensure there is a system of compensating energy production fluctuations as a result of renewables using an integrated electrical, thermal and hydrogen/gas energy network. Active demand response management, on the other hand, will provide access to consumer usage data as well as ensure the active participation of consumers to influence flexibility in demand.

a) Energy demand in buildings

An active demand side response mechanism will also cover demand side energy efficiency in order to realise significant reductions in energy consumption by considering account rebound effects. This will involve making upgrades and updates to infrastructure, as well as understanding the unique characteristics of renewable in comparison to conventional energy production.

To do this effectively requires doing the following:

- attaining a deep understanding of the variety of responses and their benefits arising from the country stock in view of the range of coordinated control approaches;
- exploring the disparity in energy service demands as a result of demographic and socio-economic characteristics.

The demand for energy services is influenced by the level of human activity. Details on the amount of time spent per household are readily available for a few countries in the Time Use Survey. Such types of details allow users to make a model that details the timing of appliance use, hot water demand and space heating.

The household activity model randomly creates a representation of household-level activities on weekdays or at weekends for a specific household size and type.

The activity profile has enabled energy savings to be optimised through the use of automated controls. One of the biggest challenges, however, is to identify activity profiles and energy demand in space and time.

b) Energy demand in cities

The importance of energy demand is particularly crucial for cities. Owing to the transition from an agriculture-based economy to a mass-market industry, the proliferation of technologies and services has led to a substantial rate of population migration from rural to urban areas.

Since 1950, many cities have witnessed a significant rise in population from rural areas, with consistent growth rates over many decades. Figure 10.1 illustrates that, from 2010 onwards, more than half of the global population are living in cities. Future estimates show that nearly 60 per cent of the global population will live in urban areas by 2030, which will be followed by a 70 per cent increase by 2050, according to the World Health Organization in 2014.

Tokyo and New York were the only two megacities in the world that consisted of a minimum of 10 million inhabitants until 1970. After this period, the number of megacities rose substantially, some of which are Delhi, Shanghai, Mexico City, Paris and Moscow. London, which has a current population size of more than 9 million inhabitants, is expected to witness a rise in population to nearly 11 million inhabitants by the year 2025.

This means that the twenty-first century will be a century of megacities, in contrast to the twentieth century being the century of urban sprawl. Asian megacities in particular are the focal point of economic development and growth, not only in terms of population size, but also in terms of the maturity of industries and other market factors.

Many cities in Asia are witnessing a considerable amount of urban growth, owing to a number of reasons. These include the forces of agglomeration, impact of large-scale economies of scale, greater employment opportunities and impact of the global economy, in addition to easier access to quality cultural, social and health services.

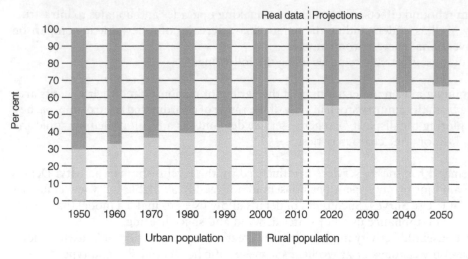

Figure 10.1 World population living in urban and rural areas. [Data source: 1,2]

This kind of growth is being witnessed in cities consisting of a combination of industrial, residential and commercial developments spread over a large area. Perhaps the best example is that of Bangkok, Thailand, in Southeast Asia.

The colossal growth in population will present new challenges in the ability to achieve a resource nexus, despite there being a considerable opportunity to achieve higher economic growth. The higher the population, the higher the demand for energy, and the more difficult it will be to utilise resources efficiently to prevent a shortage. It will put an upward pressure on the need to utilise resources, electricity and other forms of energy efficiently so that future generations will be in a position to meet their needs.

The relationship between a country's quantity of used energy and the size of its economy is a strong one. However, what makes this relationship problematic is that it is non-linear; nevertheless, the right combination of economic policies can be utilised to reduce the amount of energy needed to increase national GDP.

Taking the example of China, it has been demonstrated that it is in a position to achieve economic growth without incurring a significant rise in GHG emissions. China's CO_2 emissions dropped between 6 and 14 per cent from 1996 to 1999, while its economy expanded by more than 22 per cent. In the US, on the other hand, CO_2 emissions have increased by nearly 5 per cent according to the Clean Energy States Alliance.[3]

In terms of thermodynamics, the availability of energy resources can enable users to easily determine, for example, the quantity of food grown and cooked or how a dwelling is kept heated or cooled.

More importantly, the quality of energy is just as critical as its quantity. This is because states that are solid, liquid and gaseous vary in their capacity to offer energy services depending on their end-use devices and nature of applications. Coal, for example, can provide a greater quality in terms of energy resources than wood; oil, on the other hand, can be much more valuable than coal, while electricity is greater in quality than all solid, liquid and gaseous fuels.

Thus, urban settlements do not provide a sustainable means of accommodating the needs of a growing society. For this reason, organisations are presented with a huge burden of ensuring sustainable development in the context of urban settlements, which are witnessing a substantial rise in population growth due to the rise in megacities.

Such cities are the fuel for the economic growth of a particular region or country. Castells[4] has pointed out that megacities will continue to develop both in terms of size and being a preferred location for pursuing lifestyles by people since these are the epitome of economic, technological and social progress. Megacities thus act as important drivers of growth for their own countries and also are the focal point in the network of the generation of people who are dependent on the new economy to sustain themselves.

When the urban population rises, the demand for energy will grow accordingly.[5] There is research to suggest that, for every rise in urban population size of 1 per cent, there is a 2.2 per cent increase in energy consumption. The European Commission,[6] on the other hand, expects that more than 50 per cent of global energy demand will come from developing countries by 2030.

In a separate work, Kennedy[7] attributed the way cities function with that of the metabolic processes of an organism, calling it 'urban metabolism'. This is because the metabolism of a city consists of inputs, such as water, energy and materials, which are transformed into outputs in the form of goods and waste.

At the moment, many cities have pioneered a highly linear metabolism system, while efforts to achieve sustainable urban development could materialise in the form of a circular system,[8] in which waste is minimised and reused within the city boundaries.

On a global scale, energy demand patterns vary considerably from one city to another, across multiple countries and continents.[9] Each sector, from transportation to building and industry, is impacted by the country's degree of industrialisation, climate factors, and a range of urban structure factors.[10,11]

Cities located in developing countries show a very different picture in terms of their economic development. The I&C sector, for instance, accounts for more than half of total energy consumption requirements, particularly in cities such as Shanghai; this reflects that the Chinese economy is achieving economic growth at a very rapid pace.[12]

Electricity and oil fuels constitute the biggest contributors to the energy supply mix of the megacities Tokyo and New York. London, in stark contrast, is highly dependent on natural gas, contributing to more than half of the total energy consumption, followed by electricity and oil fuels in smaller proportions. Other cities, such as Shanghai, mostly rely on electricity and coal, although the contribution of natural gas is rising.[13]

Much of the present-day problems are found in megacities. The consistent growth in population along with other societal changes as a result of the higher migration of people from rural communities to urban areas is perhaps the most pressing challenge. This raises the requirements for food, water and energy in order to meet higher demand. Such challenges are also precipitated by factors pertaining to climate change, human well-being and the global ecosystem.

The solutions to such issues cannot be resolved without changing the perception and response for appropriate action and management. One response to efficiently manage the environment and energy problems of a city involves conducting a comprehensive analysis of past and current scenarios, formulation and execution of appropriate measures, the progress of technology development and transfer, and improvement in human resources.

Even though there has been an attempt on the part of authorities to understand the complexities and challenges at the national and city level using a thorough practical model, the efforts have not materialised in a concrete manner, owing to the limited exchange of detailed analytical data.

The different types of current urban energy models have to share common goals and address common challenges so that future opportunities can be identified and met much more easily. The complexities associated with related urban energy model systems represent the biggest challenges at the moment. Unless these are reconciled and managed, it is unlikely that an accurate model can be produced.

For a start, including more assumptions as part of the model's architecture is an effective method of approaching the problem. Models are usually constructed and tailored to represent, understand and resolve specific problems for a given audience, and not to generalise across a large section of data. Nevertheless, there have been notable efforts on the part of researchers and experts to incorporate models from varying sectors in order to address the problems and intricacies present in the urban energy systems domain.[14,15]

One in particular is the DynEMo model, which aims to showcase the responses of an energy system's essential elements, along with the technology and people.[16] According to a *Times* article, the energy system costing the least is optimised in response to a number of user constraints, from medium to long-term scenarios.[17] SynCity, in contrast, is a tool kit that has been modelled for urban energy systems and is one of the few that explains urban energy systems in a much more comprehensive manner.

However, the uncertainty, quality and availability of data are some of the major issues within urban energy systems. This is because cities are not discrete entities in which the boundaries and limits to data can be set accurately. Instead, they are open systems due to the flows of exchanges of information, resources such as energy, and people.

Data, therefore, is complex; input data usually consists of consumer behaviour or transportation, GIS maps (e.g. EEP model) and GIS databases (e.g. the EnerGIS model that evaluates the demand of energy in a given urban area).[18]

A few models, on the other hand, make use of external databases. One of these is the Modular Energy System Analysis and Planning Environment (MESAP) toolbox, which is connected to GIS databases. It has been developed to simulate and assess the demand and supply of energy and assess the impact of costs and the environment on energy requirements on a local, regional and international level.[19,20]

Models normally gather data under administrative and district contexts and only a small number of them integrate life cycle assessments that exceed city limits. More importantly, urban city data is collected by different people at different times or is adjusted using other sources either by scaling up household data or scaling down data pertaining to national energy. An example of this is the DECoRuM model, which aggregates data of CO_2 emissions obtained from individual households on an urban city-wide scale.[21]

As a result, the modeller has to decide whether the quality of the data meets the research objectives and aims in terms of accuracy and reliability. Nevertheless, improving the standards of data collection and data sharing can help resolve these challenges.

As pointed out by Keirstead, establishing an urban energy glossary can be extremely beneficial, in addition to urban energy model ontology that will consist of a database of cities' energy data; this will allow data collection and sharing to prosper on an international scale and help facilitate the ease through which accurate models can be developed.

Demand analysis is divided into multiple strands of data using a hierarchical tree structure across many levels. These can range from highly fragmented data structures to highly aggregated assessments. Usually, a structure would be broken down into sectors including transport, agriculture, industry and households, each of which will be further split into various subsectors, fuel-using devices and end-uses.

Ensuring there is a range of organised data structures gives users a considerable amount of flexibility in adjusting their structures to objectives based on how much data is available, the nature of evaluation they intend to conduct, and preferences for units. It also allows them to establish different points of aggregated data for each sector.

Although this is useful for attaining an overview of the energy demand and supply data, users still require a far more in-depth approach to ensure renewable energy data is properly incorporated into an urban energy system. Also, a city perspective within the system is highlighted to shine a spotlight on the critical technologies that can help ensure greater economic benefits and reduce environmental damage.

The benefits of estimating useful energy are twofold; it is beneficial in terms of incorporating renewable energy into an energy urban system and for enhancing learning opportunities through the use of optimisation models, such as MARKAL, which involves using demand energy as the primary driver of data in the model.

c) *Energy demand estimation*

In multiple countries, the in-depth hierarchical breakdown of a city's energy requirements cannot simply be gathered from national statistics or surveys, especially in the case of developing countries.

The data figure that is measured at the city level is mainly limited to the overall consumption of a number of energy forms. As a result, due to practical reasons, it is often beneficial to use a bottom-up approach as opposed to a top-down approach in order to obtain as much detail as possible. In the study that we are going to explore, the details of the disaggregated energy demand will be utilised to give a total figure of the demand for energy, in addition to measuring indirect energy consumption.

The reason the bottom-up approach is used is due to the fact that there is no requirement for using electricity, oil, fuel and so on. What is required is the energy services used for achieving multiple activities for each sector. Each of these activities is offered using an end-use device, such as a boiler engine that essentially converts electricity, fuel and other forms of energy into useful energy.

Using this type of approach is not only useful for providing the parameters of establishing balances in energy, but also enables users to attain an improved estimate of the dissemination of end-use activities, such as solar collectors for water heating.[22] However, energy services are intangible, particularly in their procurement of what constitutes real energy levels. Examples of this include the transportation of goods or of a person from point A to B and the provision of a well-lit work environment or air-conditioned rooms, and so on.

In this regard, what makes energy useful is that the work done to provide energy services is equivalent to its offering level. Among the many examples of these is the amount of heat produced from a room's heating system that is maintained at a specific temperature, the force used to accelerate a vehicle, or the light radiating from a lamp. The amount of useful energy can be determined by multiplying the data of end energy use with that used by an appliance.

The process of arriving at an estimate of useful energy demand can be described as follows.[23]

Space heating method

This is used in the estimation of space heating energy needs in the building sector. This approach relies on an easy-to-use model of the building's thermal heat balance. The amount of heat loss is measured using a series of equations that consider the amount of heat lost through air infiltration and the building shell. The essential parameters required for the estimation procedure are the building number and its characteristics.

Appliance saturation method

This is used in the estimation of energy demand in electrical appliances and air-conditioning. There are two parameters for executing this method: the number of households and the amount of appliance stock. The latter are further broken down into multiple sources of information, such as technical characteristics, the level of saturation per appliance type, and the mode of usage.

Floor space method

This approach is mainly used for measuring data in the tertiary sector. With the exception that end-use activities are determined on the basis of a building's cross-sectional area, the floor space method is very similar to the appliance saturation method. The electricity requirements of air-conditioning and lighting, for example, are represented according to unit of area. For the agricultural sector, on the other hand, energy demand is estimated according to cultivation areas as opposed to building areas.

Thermophysical law

In this law, data is used for estimating the energy requirements for water heating. This approach hinges on using population data and other important parameters for hot water, including water consumption per capita, number of tourists, comfort standards, or temperature of hot water.

Statistical energy data records

Statistical data for electricity consumption and fuels are normally available at the city level, particularly with the industrial and transportation sectors. This is so that demand for energy can be collected from statistical records directly. However, estimating energy demand always depends on the degree to which primary data can be found from statistical records, and whether it meets the objectives of the required data.

Energy demand that is deemed useful as according to LEAP[24] is established upon a set of projections consisting of drivers, such as GDP growth, number of buildings, population growth and the number of passenger-kilometres. In the case of the transport sector, many modes of transport such as diesel trucks, petrol cars and electric trains convert fuel into an energy service. Considering this service is far more beneficial than useful energy. The ration of conversion between the delivered service and fuel is known as intensity instead of efficiency.[25] The services delivered in the transport sector are passenger-kilometres for the passenger sector and ton-kilometres for the freight sector.

It is also important to compare energy consumption's aggregated figures with current statistical records to ensure that the model is consistent. This calibration process is needed to provide a secure basis for the energy model through which future estimates can be made with a greater degree of accuracy and reliability.

d) Non-domestic energy demand and emissions models

The literature review concerning non-domestic energy and emissions is extremely extensive. Pout (2000),[26] for instance, constructed a model known as N-DEEM, a national and non-domestic energy and emissions model that measures the amount of energy consumption patterns of the non-domestic building stock. These were categorised primarily according to activity sensors, but also on the basis of built form types, and the use of materials and technologies with respect to their age.

These extrapolations are then further combined to provide an aggregated view of the national consumption of energy using public floor area statistics. This exceptional work was also expanded upon by Bruhns and Wyatt,[27] who developed a framework of data by joining mapped information with the data gathered from multiple in-depth energy surveys.

Choudhary[28] modelled the energy consumption between buildings of the same sector of activity using Bayesian regression. Such a regression analysis technique is intended to give a clearer insight into the range of energy efficiency in the non-domestic building stock. The only parameter that is taken into account to describe the non-domestic building stock is the sector of activity.

Parameters that describe the building, such as age of the building, level of insulation and use and size of technology, for example, are not considered as model entries.

e) Demand response in the large non-energy-intensive (LNEI) sector

The demand response can facilitate a significant reduction of electricity system and operational costs, in addition to emissions and energy consumption. The absorption of variable renewable electricity, along with dispatchable fossil fuel and nuclear generation efficiencies, can be increased by enhancing the demand in time and quantity levels.

To achieve this, demand can be adjusted by rescheduling it with storage factors, such as service, chemicals and heat. This can be done either through dual fuelling, which involves switching from electric to gas heating, or through service interruption. The significance of demand management is only expected to grow with the electrification of end-user supply systems, electric vehicles, an increase in electric heating, and the rise in the renewable fraction of supply, particularly wind power in the UK.

The UK government, in its 2009 report, pointed out that the commercial sector has the biggest potential for discretionary demand, which is far more than the domestic or SME sector or the energy-intensive industry.

The data and modelling at UCL in the UK indicate that this view may not be accurate. The most easily load-managed demand is for heat, as storage is practical and relatively low cost. To understand the load management details in different sectors, it is necessary to model all sectors, since the load management in one sector partially depends on the whole system.

The main difficulty is the lack of publicly available data on the detail of sectoral and subsectoral demands and their variation over periods of an hour or less. The UCL modelling work uses an electricity services model, *EleServe*, which includes a detailed demand module, a dynamic generation module and a load management module that alters demand to reduce short-run avoidable generation costs.

Let us take the example of Greater London, UK. The city of London consists of 33 Local Authorities. These are called boroughs (with the exception of the City of London, which is still a council) according to London Councils (2014).[29]

More than half of the non-domestic building stock is represented by educational buildings (schools, colleges, universities), followed by offices (including government, private sector and courts), community halls, centres and religious buildings.

Choudhary in 2011 analysed the non-domestic building stock using building activity as his research's primary classification. He developed an alternative aggregated classification to the SIC comprising 11 primary sector categories (office, retail, primary care, hospitals, education, hotels, sports, culture, community, industrial, other) and regrouping subcategories that have the same energy consumption specificities for Greater London as follows:

- office: government, private-sector, courts;
- retail: high street department stores, centres;
- primary care: health centres, clinics, surgeries;
- hospitals: hospitals, medical research, nursing homes;
- education: schools, colleges, universities;
- hotels: all hotels and boarding houses;
- sports: gymnasium pools, leisure centres, sport centres;
- culture: cinema, theatre, performance halls, museums, galleries, clubs;
- community: halls, emergency services, religious buildings, community centres;
- industrial: transport terminals, factories, warehouses, storage;
- other: agriculture, unused buildings, freight handling.

f) De-carbonisation of heat and cooling

Cooling and heating decarbonisation are important issues when it comes to meeting the challenges of GHG reduction. De-coupling energy vectors is inefficient for both operation and planning. There is an insufficient understanding of multi-energy issues and modelling tools.

Distributed Multi-Generation (DMG) can increase production efficiency exponentially. Multi-generation as well as integrated energy networks can result in low carbon emissions. This system has a structure that is comprised of a set of multiple generation units that are supplied by appropriate fields and connected to each network of the energy-based network.

The main brick is cogeneration. Also known as CHP, it refers to concurrent electricity and heat generation from a fuel source. However, the effectiveness of the process is reliant on good environmental performance as compared to the separate production (SP) of heat as well as electricity.

A CHP system makes heat recovery possible, which would be completely discarded otherwise during the thermodynamic cycle. This means that about 80–90 per cent of energy efficiency and carbon dioxide emissions are also reduced depending on the fuel that is used. There is a need for an appropriate number of thermal users that have consistent thermal requirements for an entire year. Heat networks that make use of a number of users can increase the deployment of CHP.

10.2 Energy supply

Carriers as well as energy sources that offer energy-based services should ensure a supply that can last long term, remain financially viable and have a minimum carbon footprint. On the other hand, these three goals can clash. Currently there are enough energy sources available to provide consistent energy for at least a couple of decades even with energy sources that offer high energy conversions. How these resources can be used without harming the environment and provide for an increasingly growing populace will be a challenge in the future.

- Current oil reserves will peak along with natural gas reserves. What is uncertain is when and what will be the transitions to other liquid-based fuels such as biofuels, oil shale, heavy oils, coal to liquids, etc. However, there is no confirmation about how these alternative fuel sources will populate the market and what changes they will induce in worldwide GHG emissions.
- Traditional natural gas reserves provide more energy compared to oil, but they are also not distributed as much compared to the latter. That is not to say that unconventional gas resources are not abundant, but there is uncertainty regarding future development and its role in future economics.
- Coal is not distributed as consistently and evenly but it is still abundant. It is easily convertible to liquids, gases, heat and power, but for more intense utilisations, it will need feasible CCS technologies if the GHG emissions it generates are to be limited.
- A popular trend involves energy carriers that offer increased efficiency as well as convenience, and keeping away from solid fuels in favour of liquid and gaseous fuels and electricity.
- Nuclear energy can be utilised to provide electricity that has no carbon footprint. It comprises 7 per cent of total primary energy, which makes it a viable source of heat and energy. The main obstacles include long-term fuel resource limits and recycling, economics, security, waste management and propagation, along with unfavourable public opinion.
- Energy sources that can be renewed except for large hydro are distributed on a wide scale as compared to fossil fuels. The latter are abundant at single locations and need to be distributed. In other words, there is a need to use renewable energy in a dispersed or concentrated manner in order to meet the high energy demands of cities and large industries.
- Renewable energy-producing technologies that are non-hydro such as solar, wind, geothermal and biomass do not provide a sufficient global heat and electricity supply, but they are increasing on a rapid scale. The only factors that are preventing their growth are finances and environmental concerns. In other words, unless there are supportive government policies backing them, these energy resources may not be used to their full potential.

- Even though the typical biomass that is used for heating and cooking makes up more than 10 per cent of the global energy supply, it could be replaced in the future by contemporary biomass and other renewable energy systems. Other alternatives could include fossil-based domestic fuels such as kerosene and liquid petroleum gas (LPG) (high agreement and much evidence – except traditional biomass).

If future energy-supply constraints and disruptions are not met with viable solutions, complications are imminent. At the present time, fossil fuels are used to provide approximately 80 per cent of the world's energy supply. This is separate from their conventional use as zero- and low-carbon-emitting energy systems including detaining and storage of carbon dioxide (CCS)[30] along with better energy efficiency, which would be part of the solutions when it comes to GHG-emission reduction.

However, there are still some uncertainties when it comes to determining the technologies that can assist this transition and the policies that can provide relevant momentum, although energy supply security meets GHG-reduction goals, both of which act as co-policy drivers for governments that want to make sure that their future generations will be able to provide for themselves without compromising energy sources.

Energy-based GHG emissions are actually by-products that are produced by the conversion along with the delivery sector. This includes refining, extraction, transference of energy carriers via ships, pipelines, etc., along with sectors that rely on energy such as transportation, waste, forestry, buildings, agriculture, etc.

Incorporating major energy transitions will not be a quick process. The diffusion rate of new energy technologies is dependent on the expected years that the capital stock will last along with the cost and the equipment needed. However, certain energy conversion plants can last for more than a century with a slow turnover rate. An approximate 2–3 per cent replacement rate for a year is quite common.

The conversion from abundant fossil fuel resources to restricted gas and oil carriers, along with a new energy supply and conversion technologies, has started. The only thing keeping it from progressing is regulations and general acceptance; rapid implementation and market performance alone may not reduce GHG emissions on their own.

There are a number of nations with energy-based systems that have transformed and evolved from their reliance on fossil fuels due to climate threats, increasing dependence on energy markets across the globe and the failure of the supply chain, all of which have contributed to wiser energy usage.

In other words, the world is not on the right path when it comes to securing a future of sustainable energy resources. Fossil fuels will continue to dominate the world for the next several decades, but zero- and low-carbon technologies will need to be utilised to reduce GHG emissions. This is possible as business opportunities and co-benefits emerge, but rapid deployments of zero- or low-carbon technologies will need new policies for: security of energy supply; removal of structural advantages for fossil fuels; reducing environmental impacts that can result from these efforts; and acquiring sustainable developments.

10.3 Energy storage

Storing energy is a cheaper and more cost-effective solution when it comes to system-level flexibility across:

- time: seconds to months;
- space: domestic to generation-integrated;
- vectors: heat and power.

Launching an energy storage system requires the interplay of applications as well as a different technologies system value vs market opportunity, along with short-term costs vs long-term benefits. To make the employment of storage technologies possible, the dynamic of an energy system transition is needed, the main reason being that it is likely that intermittent generation will expand before a demand response from electrical vehicles (EVs) and heat pumps (HPs).

There will be an increase in demand for electricity for heating and transport in the late 2020s, and challenges will become more serious by 2050.

a) Making the case for storage

With an increase in generation from continuously operational and discontinuous sources, along with a rising demand for unpredictable electricity with less predictable profiles, a flexible energy system is a must.

Energy storage can capture off-peak or excess generation and deliver at peak times without compromising the national security of the supply. In addition, consumers do not have to change the way they interact with technologies to use them.

Some of the barriers to deployment include:

- *technology cost and performance*: other technologies are currently more cost effective in comparison;
- *uncertainty of value*: the future value is dependent on the energy system mix;
- *business*: capturing multiple revenue streams is difficult to establish, both for a potential business and the market in which it will operate;
- *markets*: the true value of energy is not reflected in the price; the future long-term value of storage cannot be recognised in today's market;
- *regulatory/policy framework*: restrictions on ownership. This includes paying levies twice;
- *societal*: wider community acceptance is yet to be considered.

b) Value of flexibility

'System flexibility should be considered as a critical factor for designing decarbonised electricity systems and facilitating cost-effective evolution to lower carbon electricity system.'[31]

For example, Strbac *et al.* draw several key findings for the UK. The values tend to be higher than previous studies suggest. In the scenarios with high renewables the value of storage increases markedly towards 2030 and further towards 2050; a few hours of storage are sufficient to reduce peak demand and thus capture significant value; and storage has a consistently high value across a wide range of cases that include interconnection and flexible generation.

However, the 'split benefits' of storage pose significant challenges for policymakers when it comes to the development of relevant market mechanisms to ensure that the investors in storage are adequately rewarded for delivering these diverse and valuable sources.

c) Linking energy systems enables multiple storage use

Loads can be shifted with the storage of:

- heat
- hot water tank, bricks, etc.
- cooling in a refrigerator
- service (e.g. running the dishwasher at night)
- other energy sources such as a powered battery.

The major potential lies in heat, and this is already widely deployed in ~15M hot water tanks and ~1.5M off-peak electric storage heaters. However, there are some difficulties that can be brought about by the loss of heat and the impact of storage temperature on the performance of the heat pump; the higher the temperature of the tank, the lower the coefficient of performance (COP) of the heat pump.

There are a number of priorities for innovation in energy storage, from the *analysis* of the value of energy storage and other flexible options in the energy system during the transition period, to *demonstrations* and understanding of results at the distributed level and an *R&D effort* in order to develop lower-cost alternatives.

d) Conclusions

Summarising what we have learned:

- Energy storage could play an important role in a cost-effective transition to a low carbon, resilient energy system on a global scale.
- Technologies currently exist that can respond to energy system challenges, as their value will increase once they start reducing costs.
- However, short-term fixes could crowd the market for more efficient long-term solutions.
- Support needs to be well coordinated and strategic, across the innovation process, and this includes research on deployment.
- The whole system, i.e. electricity/heat/cold, will be considered.
- Innovative policy and regulation have to provide a market through which value (of flexibility) can be accessed.
- Energy storage can raise the prospect of new business models for energy.

10.4 Power-to-gas

In the latest 100 per cent renewable energy scenario of the Federal Environmental Agency (FEA), power-to-gas (PtG) is the most important balancing and storage technology.[32] Using a system perspective, the system predicts a PtG capacity of 44GW electricity input in order to balance the energy system. In the FEA scenario the plants then only operate when surplus electricity is available mainly from wind energy.[33] This then begs the question – why should anybody invest 1,750€/kWel input in a power-to-gas utility and operate it in an intermittency mode, only in peak wind hours?

The long-term perspective is based on a description of the recently published FEA 100 per cent scenario, i.e. the initial scenario that considers PtG for a large-scale integration of renewable energy. It basically concludes that a PtG capacity of 44GWel input will be necessary to cover storage demand until 2050 and absorb the production of renewable energy sources. After that the scenario is compared to other scenarios and solutions on a critical level and towards a 100 per cent renewable energy system.

a) Power-to-gas process concept

The concept of power-to-gas relies on a technology that is as old as electricity itself. On 20 November 2008 Gregor Waldstein filed a patent application in more than 15 countries for the 'Modular Power Plant Unconnected to the Grid'. This entails at least one system of electricity production from sustainable sources that supply power to the following subsystems: a CO_2 desorption system; an electrochemical or solar-thermal dihydrogen synthesis system; a catalytic synthesis system for methanol or dimethyl; and a storage system for the gas synthesised. The different subsystems need to be settled into capacitance with each other. The patent is what gives the inventor or the assignee the right to make use of this process during the time of the validity of the patent – that is, until the year 2024.

Electrolysis

This is the process of pushing chemical reactions with the use of an electric current. This was first used by Martinus van Marumto to produce antimony, zinc and tin. Fifteen years on from this, William Nicholson, an English chemist, along with Anthony Carlisle, a surgeon, found out how electrolysis with water mixed with a new voltaic pile can make O_2 and H_2. In the year 1869 it was the Gramme machine that eased the use of electrolysis with water; in 1888 Lachinov created an industrial process to forge oxygen and hydrogen through electrolysis.

In the twentieth century the whole process of water electrolysis was worked and improved on, with the whole process diversifying at the same time too. From this, three main techniques emerged. These were high temperature electrolysis, electrolysis and polymer electrolyte membrane (PEM). Electrolysis of alkaline water is known for its cheap technology. It was invented at the start of the nineteenth century, and now alkaline electrolyte is the most dependable kind.[34] Their electrical efficiency has only reached 60 per cent. PEM electrolysis has been a lot less efficient recently than the spread technology.[35] PEM is considered 75 per cent efficient but is still more costly. The main benefit to be had here is the adaptation to current variations which are particularly suitable for solar electricity or wind supply. That being said, their power is limited; the relative newness of this technology prevents its use at an industrial level.

Solid electrolyte electrolysers need a temperature of 700 degrees to operate. This cuts down on electricity consumption and no longer makes it important that the catalyser be used. The recent technique sounds promising. The three electrolysis technologies have benefits, but the thing that leads to the debate here is the final price of the hydrogen that is produced. The oscillating price falls between 10 and 5 euros for each kilogram of hydrogen that is made, which of course depends on the electricity price. The French Alternative Energies and Atomic Energy Commission (CEA) claims that solid electrolysers will lower the cost of this to under 3 euros for each kilogram in a couple of years.[36]

Methanation

When H_2 is generated, it can be used for methane synthesising. The process is referred to as methanation; in opposition to bio-methanisation it is also referred to as anaerobic digestion and methanisation that creates CH_4, only it makes use of biodegradable substances. CO_2 methanation is the most interesting reaction here, especially because it can cut down on CO_2 emissions on a very large scale with CO_2 taken from industrial sources. The technology is very recent and is still subject to continuous development, most importantly in the area that discovered the top catalysts.

The process of methanation has been utilised in industries since the 1970s, especially in the production of synthesising natural gas from coal. The reactors that are currently used are ones such as fixed-bed and fluidised-bed reactors.[37]

Latest studies are looking into the process of a three-phase reactor that can perform methanation and improve heat management. In both systems it is the reactants that are gaseous and the catalysts that are solid. On top of this, the three-phase systems come with a liquid heat transfer medium which is never required for any two-phase mechanisms because of reactor configuration.[38] Even with two-phase systems being commercialised, it is the three-phase systems termed bubble column that are currently being tested and developed.[39]

The main benefit of a three-phase system is that it can handle a lot more load variations than a two-phase system can.[40] Industrialising the three-phase system for methanation is going to be a major step ahead for PtG development. Methanation always follows the creation of H_2 in all PtG processes. CO_2 is also required.

b) Storage, transport and distribution

Like the layout of the PtG procedure portrays, methanation might not be the only resulting possibility. Hydrogen has always been used as the final product, but that does not mean that we don't compare it to CH_4 – most notably in terms of transport, distribution and for storage. When there are normal conditions of temperature and pressure, H_2 is never dense. Because of that, storage becomes the only option when influenced by specific conditions via the three methods of extreme low temperature liquid storage, solid hydrogen containment, and high pressure containment.[41] The methods of high pressure and liquid containment require energy, which is a major drawback, since the aim here is to store energy and not consume it. Thanks to the very last method that was developed by McPhy, it is possible to store 100 kilograms of hydrogen every cubic metre. That being said, it also has the disadvantage of being heavy, and because of that is relevant only in the scenario of stationary containment.[42]

It is physically possible to add H_2 to the natural gas network using up to 50 per cent of total capacity, but safety limits it to 30 per cent. User appliances use hydrogen in a proportion ranging from 0 to 20 per cent.[43]

In case hydrogen is ever used for natural gas synthesis, utilising the existing network for this has never been an issue. The only requirement here is that a

PtG plant should be connected to a grid that is able to insert CH_4 inside it; the pace of this needs to be just as fast as the time taken to create it. In this scenario the network acts as a containment system because it is great at supporting load variations.

For example, in the UK, Energy UK evaluated that a gas surplus is shot to storage facilities such as salt caverns and diminished gas fields. Either that or it is liquefied to be stored in LNG facilities. Once these are filled to the brim, they only manage to meet half of the entire United Kingdom's yearly demand.[44]

The main objective of going through the whole PtG procedure is so that one can store large amounts of energy that would otherwise be simply lost. Hydrogen and synthetic natural gas are what make room for the huge energy containment that can go up to TWh scale. The discharge time here is adapted to that of multiple-day containment. CH_4 and H_2 containment power was never the highest among many different stationary storages, but if used as a medium-term energy containment it could work as a great option.[45] This is because storage and network facilities exist and do not need any major changes.

Natural gas containment, transportation and the distribution in its synthetic form is therefore a valuable asset of PtG concept technology. The ability of the system is to store hydrogen, notably controlled by user regulations. The National Grid, being one of the essential gas distribution companies, is currently taking up work that may widen the scope of the use of H_2.

c) Efficiency and costs

The science of converting energy has always been the foundation of our scientific society. The efficiency of this conversion is an important factor when you choose an energy source or a containment method. In the case of the PtG process, note that it is extremely important to keep in mind that, when you are choosing a containment method or an energy source, you keep track of its efficiency. In the case of the PtG method, remember that electricity is used in excess, and because of that the efficiency ratio does not have a similar effect to that in other industries.[46] That being said, you should not always aim to maximise this ratio. Another thing to be taken into account is cost.

Current and forecasted efficiency factor

Currently there are three separate efficiency factors that you must consider here; each of these helps to be one step ahead in the PtG process. The first factor here is the conversion efficiency of water electrolysis. The speed of the process and the voltage is what determines the efficiency. The next move in the cycle is methanation. The efficiency of power in the methane gas process is placed at around 65 per cent. This cycle is entirely completed when synthesised methane gas helps generate power. The company SIA Partners thinks that generating electricity out of methane comes with an efficiency of 77 per cent. Electrolysis is the method that leads to major losses with plenty of different technologies. The three dominant technologies established earlier come with their own range of use and efficiencies. It is clear that future efficiency is going to surpass what we are seeing at the present time. Investors may favour this and assist in the development of the industry.

Costs of the power-to-gas process

Several researches have been done to analyse the economics of the PtG process. In 2012 Kolberg modelled the costs of a 20MW PtG facility that was to be located in Ireland.[47] Some assumptions made in the work are:

- the Balance of Plant (BoP) system costs 1,000€/kWel, including 765€/kWel for industrial solid oxide fuel cells (SOFC);[48]
- electricity costs 0.03€/kWel due to the low demand;
- use of the rainwater harvesting system for water collection and include cost in the fixed costs;
- CO_2 is provided at no cost by the source.

If the cost of electricity of 3c€/kWh is an investment beyond 10 years' time for an interest rate of 5 per cent, the maintenance and operation costs will be 3 per cent of capital cost, the total being 87,600MWh power input for each year. The factory produces 47,500MWh of SNG annually, which is the cost of €0.11 per kWh. The cost varies from 4 to 11c€/kWh, depending on payback time and the electricity price.

The cost is then compared to the cost of imported natural gas. The United Kingdom Energy Saving Trust estimates the average domestic price for this to be around 5.33 c€/kWh or 4.21 pence/kWh. The commercial price on average is then 3.92c€/kWh or 3.1 pence/kWh. An interesting thing to note here is that improvement of technology lowers the cost of balance for plant systems to an extent where SNG becomes the alternative imported gas.[49]

d) *Successful examples – Germany*

In Germany there are many actors in the energy market who have already started experimenting with PtG methods. This comes as no surprise because the German government likes to take strong measures when it comes to the generation of electricity.

Living next to a power plant is not an issue for 65 per cent of the German population; in the UK only 9 per cent accept living next to a coal-fired power station, while 3 per cent don't mind living near a nuclear plant. Around 80 per cent of Germans believe that renewable energies that fall to around 3.5 euro cents for every kWh are way too low.

e) *Ongoing projects*

Germany may be the most modern-thinking nation in the field of PtG, but there are other countries that have funded such projects. ForskEL, the Danish research company, invested €3.7 million in a process called BioCat.[50]

This process is soon to be installed in some of the biggest wastewater plants in Denmark. It will consist of a 1MW electrolysis device with a bioreactor where hydrogen and carbon dioxide will be mixed to make renewable methane. This process will be used to store part of the renewable electricity that will be generated by significant wind power turbines.

f) Power-to-gas – key figures for economic modelling

At a demo scale, investment cost is 2,000€/kW (15,000kr/kW). The cost covers control systems, civil construction, piping, power electronics, compression, civil work, methanation and electrolysers. The system efficiency power of methane stands today at 60 per cent.

Future investment cost is 200MW and may decrease to 1,000€/kW (7,500kr/kW) with the use of SOEC-efficiency at 65–68 per cent, or maybe 75–80 per cent. Market prices for natural gas are:

- Nordpool price level, October 2012 – 26€/MWh or 2.2kr/m³ methane;
- feed-in tariffs: feed-in tariffs for 'green gas' – 115kr/GJ or 4.56kr/m³ methane;
- per October 2012 – selling price for 'green gas' to the Natural Gas Grid might be: 2.2 + 4.56 = 6.76kr/m³ methane.

g) Conclusions and future research

Power-to-gas may just be the solution for long and short-term containment problems posed by fluctuating energy. Its benefits are:

- cutting down the exploitation of fossil fuel resources;
- contributing to less import dependency;
- raising energy security;
- improving the exploitation of renewables;
- making sure that the base requirements for wind power are met;
- enhancing the incorporation of renewable energy into the National Grid.

More research is needed for fuel cell technology and related processes. Solid oxide cells allow for the reversible flow of gas and electricity in the form of SOFC and SOEC modes as the greatest potential for large-scale applications without there being a need for conventional combustion turbines.

There is a need for more research in the area, including: analysis of carbon rates and credits for PtG storage, and investigation into the effects of burden sharing of the load balancing cost by non-synchronous power plants.

10.5 Energy: centralisation versus decentralisation

Electricity production in Europe was driven towards centralisation due to the development of large power plants by the end of the postwar period. The electricity system in Europe is undergoing continual transformation. Currently, Europe is experiencing a period of modifications in the energy sector, which is the reason why reliability, energy efficiency and a decrease in greenhouse gases have become a priority where the sustainable energy strategy is concerned.

The EU legislative field has not specified the questions that are posed by the decentralised approach, which leads us to Donald Rumsfeld's 'unknown unknowns', according to the definition of decentralised energy as given by the DTI: 'power generation that is connected to the low voltage distribution network at 132kV and below'. We can say that the benefits of decentralised energy could be at a low level; because of this it can contribute towards an established distributed generation as a main source of electricity or as a backup producer, with decentralised energy being generated in smaller power plants.[51]

Currently, 93 per cent of electricity worldwide is supplied through centralised generation and distribution.[52] However, around 65 per cent of the energy is lost before it even reaches consumers.[53] If this wasted heat can be used, it can contribute to improving energy supply security and tackling climate change.

The current investments are focusing on centralised technologies especially in electricity because of the structure of the electricity and gas markets, the way they are regulated, and the fact that much industry and many policymakers support the convenience of centralised energy systems.[54]

Energy consumers can become producers of energy, with distributed storage and generation, which is why emphasis is laid on the point that a decentralisation system has several advantages over a centralised system. Decentralised systems can be tailored to local conditions, which could revolutionise the lives of billions of people who currently lack access to basic, clean and affordable energy services.[55]

Despite the benefits, there are still significant barriers. It is important to note that decentralisation has already started to make a strong appearance in the European energy market.[56] Take the example of EU members such as Denmark and the Netherlands, where 50 per cent of electricity and almost 40 per cent of energy, respectively, are the result of decentralisation. Other countries such as Sweden, Germany, Austria, Finland, Italy and Spain are undergoing such a change.

10.6 The need for a 'whole' integrated approach

An integrated 'whole' system approach in space and time at different levels is needed to understand the energy delivery strategy for different geographical zones.[57] The tool(s) should principally configure to provide the lowest lifetime cost solution that is consistent with the constraints applicable but should also facilitate an examination of multiple results, in order to allow the assessment of more costly, but potentially relevant, solutions. This can include solutions that deliver the lowest net CO_2 emissions, are the most resilient or best satisfy a metric related to energy provision or fuel poverty. Alternatively, the proposed solutions may highlight additional potential issues that were not initially considered. So there may be a desire to investigate alternative options that may restrict the deployment of certain types of technologies based on geography, time frames or overall resource capacity.

It should be noted that there is no requirement for a single tool to be developed in order to fulfil all of the requirements, but there can be a set of integrated tools that can offer an appropriate solution. It should be noted that references to 'tool' as opposed to 'tools' in this book should not be seen to preclude the option that a set of integrated tools may be an appropriate solution.

Such a 'tool' should continue to build on existing work and should be applicable at the whole country level, taking into account the building types, local geography, occupancy types and preferences, energy infrastructure implications, local climate conditions, costs, performance and availability of the range of component technologies that might form part of the future system.

The intended outcome would be to understand what component technologies predominate under the range of different scenarios, conditions and operational strategies.

10.7 Scenarios: results vs reality

It is important to consider different scenarios in order to analyse possible paths and implications over time. As an exercise we looked at previous scenario studies that were published in 2006–2007 that made worldwide predictions for 2010 and then compared their results with the actual data. The criteria for inclusion were to be worldwide in scope, backcasting and comprehensive. The importance of assessing these historic energy system projections could help us to improve future energy scenarios.

The studies considered are:

* *World Energy Outlook*, International Energy Agency (IEA), 2006;[58]
* *Energy Technology Perspectives*, IEA, 2006;[59]
* *Energy [r]evolution*, Greenpeace and EREC, 2007;[60]
* *International Energy Outlook*, U.S. Department of Energy, 2006.[61]

The actual data source for 2010 was taken from World Bank and Enerdata. The following comparisons were made:

Population

Population is included in all of the energy reports as a common assumption among the different scenarios. Population and demographic data are the key indicators that have an impact on global energy demand. In the studies used, the annual growth of population is assumed separately for each region but a common growth factor is applied at a global level. The annual growth ranged between 1 and 1.2 per cent from 2003 to 2010, but this growth factor was extended up to 2030 in the *Energy Technology Perspectives* (ETP) study by IEA. The data for 2010 was slightly higher than the assumptions in the four key energy studies that show the tendency for population expansion in the near future, especially in the non-OECD territories which include China, India, Africa and Latin America (Figure 10.2).

GDP growth

GDP growth is a common assumption in all the scenarios included in the energy studies except for the *International Energy Outlook* (2006). It is mentioned in the *Energy [r]evolution* study by Greenpeace and EREC that a 0.1 per cent increase in GDP leads to a 0.2 per cent increase in energy demand. The areas projected to present the highest development are China, Africa and Latin America, and also some Asian countries that follow the same trends with population increase. The Low Growth Scenario of the *International Energy Outlook* (IEO) study with 3.1 per cent annual growth rate is observed to be the closest with the actual rates of 2.7 per cent (Figure 10.3).

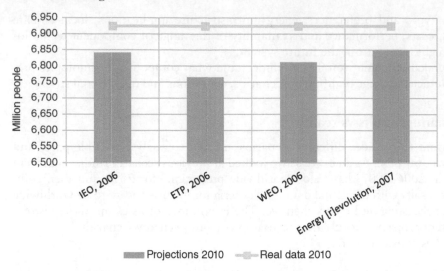

Figure 10.2 Population projections vs actual data, 2010. Data source: 62

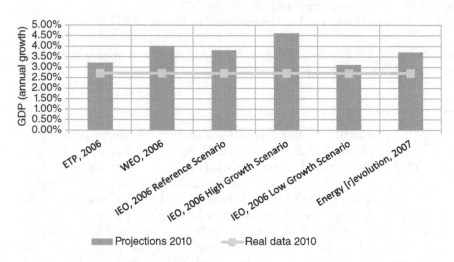

Figure 10.3 GDP growth projections vs actual data, 2010. Data source: 63

Primary energy demand

Primary energy demand is the key output of the energy scenario studies. Most studies assume that there will be an energy demand growth of 1.4 per cent (±0.3) per year by 2010. This increase is much less than that of the previous decade, which had mean rates equal to 1.8 per cent according to the *World Energy Outlook* (WEO) by IEA. The most precise scenarios come from the WEO reference case and the IEO reference case, while Greenpeace projects have much lower values based on assumptions towards an energy efficiency boost and firm climate change policies. Furthermore, it has been seen that a high energy demand increase is observed mainly in the regions that have a high population and GDP increase (Figure 10.4).

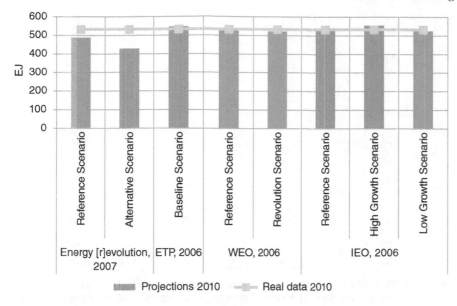

Figure 10.4 Primary energy demand projections vs actual data, 2010. Data source: 64

Primary energy mix

The primary energy mix is formed from oil and coal reductions and their replacement with natural gas as a less polluting source of energy (Figure 10.5). According to the scenarios, there is a slight decrease from 2003. Since 2010, renewable energy has been the fastest-growing source of energy and will continue to grow at the same momentum until at least 2050.

There are studies such as *Energy [r]evolution* from Greenpeace/EREC and WEO that report positive values for renewable expansions. It is seen that the IEO Low Growth Scenario shows the closest values where renewables are concerned. Only the WEO Revolution Scenario coincides with oil being significantly reduced, while it exceeded projections. Coal, on the other hand, remained at high levels in spite of the projections for a rapid decrease (a realistic result regarding this was only shown by WEO). Natural gas is growing, but none of the studies have data that shows the statistics. Lastly, WEO shows success in the reduction of nuclear power.

Uncertainties and conclusions: During this study, several limitations were identified related to the assumptions incorporated into the energy models. This is mainly because the calculation methods are not publicly available.

Also, a significant drawback regarding the projection of future energy scenarios is related to unexpected financial events. These trends are characterised by comparatively high forecast uncertainties, especially for long-term projections. For example, the financial crisis in 2008 led to a noteworthy decline in energy demand. Other factors also may have impacted on the results. Together this may have rendered several input assumptions and produced inaccurate results. In future scenarios, unexpected economic and political events should be taken into account in the development of the scenarios.

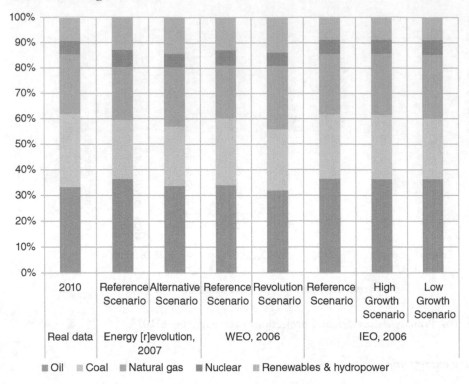

Figure 10.5 Primary energy mix projections vs actual data, 2010. ^{Data source: 65}

Notes

1 United Nations (2012). *World Urbanization Prospects: The 2011 Revision*, CD-ROM edition. Department of Economic and Social Affairs, Population Division.
2 GEOHIVE (2014). *World 1950–2050 by Continent*. Accessed March 2014 from www. geohive.com/earth/his_history1.aspx
3 U.S. Energy-Related Carbon Dioxide Emissions (2014). Release date November 23, 2015. Accessed January 2016 from www.eia.gov/environment/emissions/carbon
4 M. Castells (1993). *Why the Mega-Cities Focus? Mega-Cities in the New World Disorder*. Mega-Cities 7th Annual Coordinators Meeting, 1–7 August, Jakarta, Indonesia.
5 Urban Times (2014). *Electric Cities: The Future of Energy in an Urban World*. Accessed July 2015 from http://urbantimes.co/2014/04/electric-cities-the-future-of-energy-in-an-urban-world
6 European Commission (2003). *World Energy, Technology and Climate Policy Outlook 2030 – WETO*. Luxembourg: Office for Official Publications of the European Communities.
7 C. Kennedy, S. Pincetl and P. Bunje (2010). The study of urban metabolism and its applications to urban planning and design. *Environmental Pollution*, 1–9.
8 UN-HABITAT (2008). *State of the World's Cities 2008/2009: Harmonious Cities*.
9 J. Keirstead (2010). *Identifying Lessons for Energy-Efficient Cities Using an Integrated Urban Energy Systems Model*. London: Imperial College.
10 Eco-City Notes (2012). *Evaluating Eco-Cities*. Accessed January 2014 from http:// cargocollective.com/chinabuildsgreen/5-30-12-Evaluating-Eco-Cities
11 S. Dhakal (2004). *Urban Energy Use and Greenhouse Gas Emissions in Asian Mega-Cities: Policies for a Sustainable Future*. Urban Environmental Management Project Institute for Global Environmental Strategies.
12 UN-HABITAT (2008). *State of the World's Cities 2008/2009: Harmonious Cities*.

13 J. Ruet, F. Vallantin, A. Daval and J. Pasternak (2010). *WEC 'Energy for Megacities' Study. Shanghai Municipality Case Study: GBP 2010.* IGBP's *Global Change Magazine*, Issue 74.

14 J. Keirstead, N. Samsatli and N. Shah (2009). *Syncity: An Integrated Tool Kit for Urban Energy Systems Modelling.* Proceedings of the 5th Urban Research Symposium, Marseille.

15 J. Keirstead, M. Jennings and A. Sivakumar (2012). A review of urban energy system models: Approaches, challenges and opportunities. *Renewable and Sustainable Energy Reviews*, 16, 3847–3866.

16 UCL Energy Institute (2013). *DynEMo – UK Dynamic Energy Systems Model.* Accessed January 2016 from http://www.ucl.ac.uk/energy-models/models/dynemo

17 IEA–ETSAP (2011). *Overview of Times Modelling Tool.* Accessed March 2014 from http://iea-etsap.org/index.php/etsap-tools/model-generators/times

18 T. B. Johansson, N. Nakicenovic, A. Patwardhan and L. Gomez-Echeverri (eds) (2012). *Global Energy Assessment: Toward a Sustainable Future.* Accessed March 2014 from www.iiasa.ac.at/web/home/research/Flagship-Projects/Global-Energy-Assessment/GEA-Summary-web.pdf

19 L. Girardin, F. Marechal, M. Dubuis, N. Calame-Darbellay and D. Favrat (2010). EnerGis: A geographical information based system for the evaluation of integrated energy conversion systems in urban areas. *Energy*, 35, 830–840.

20 B. Möller (2003). *Geographical Information Systems for Energy Planning – Integration of Geographical Methods in Local and National Energy Systems Analysis*; ENERGYPLAN (2014). *MesapPlaNet.* Department of Development and Planning, Aalborg University. Accessed March 2014 from www.energyplan.eu/mesap-planet

21 ENERGYCITY (2014). *Existing Urban Energy Models.* Accessed January 2015 from www.energy-city2013.eu/pages/results/data-catalogue/existing-urban-energy-models.php

22 D. Finon and B. Lapillone (1983). Long-term forecasting of energy demand in the developing countries. *European Journal of Operational Research*, 13 (1), 12–28.

23 Y. Sarafidis, D. Diakoulaki, L. Papayannakis and A. Zervos (1999). A regional planning approach for the promotion of renewable energies. *Renewable Energy*, 18, 317–330.

24 C. Heaps (2002). *Integrated Energy-Environment Modeling and LEAP.* Stockholm Environment Institute and Tellus Institute, Boston.

25 M. I. Howells, A. R. Kenny and M. Solomon (2002). *Energy Outlook 2002: Modelling Energy in South Africa.* Energy Research Institute, Department of Mechanical Engineering, University of Cape Town.

26 C. H. Pout (2000). N-DEEM: The national nondomestic buildings energy and emissions model. *Environment Planning*, 27 (5), 721–732.

27 H. Bruhns and P. Wyatt (2011). A data framework for measuring the energy consumption of the non-domestic building stock. *Building Research and Information*, 39 (3), 211–226.

28 R. Choudhary (2012). Energy analysis of the non-domestic building stock of Greater London. *Building and Environment*, 51, 243–254.

29 London Councils (2014). *List of Inner/Outer London Boroughs.* Accessed January 2015 from www.londoncouncils.gov.uk/node/1938

30 Working Group III of the Intergovernmental Panel on Climate Change (2005). *IPCC Special Report on Carbon Dioxide Capture and Storage.* Cambridge and New York: Cambridge University Press. Accessed March 2014 from www.ipcc.ch/pdf/assessment-report/ar4/wg3/ar4-wg3-chapter4.pdf

31 Imperial UK (2015). *Value of Flexibility in a Decarbonised Grid and System Externalities of Low-Carbon Generation Technologies.* Accessed January 2016 from www.theccc.org.uk/publication/value-of-flexibility-in-a-decarbonised-grid-and-system-externalities-of-low-carbon-generation-technologies

32 Federal Environmental Agency (FEA) (2010). *Energieziel 2050: 100% Strom aus erneuerbaren Quellen.* Accessed March 2012 from www.uba.de/uba-info-medien/3997.html

33 M. Sterner (2009). Bioenergy and renewable power methane in integrated 100% renewable energy systems: Limiting global warming by transforming energy systems. In J. Schmid (ed.), *Renewable Energies and Energy Efficiency.* Accessed January 2015 from www.upress.uni-kassel.de/kata-log/abstract.php?

34 L. Kalinowski and J.-M. Pastor (2013). *L'hydrogène: vecteur de la transition énergetique.* Accessed March 2014 from www.assemblee-nationale.fr/14/rap-off/i1672.asp

35 Ibid.
36 Ibid.
37 M. Götz (2013). Evaluation of organic and ionic liquids for three-phase methanation and biogas purification processes. *Energy Fuels*, 27 (8), 4705–4716.
38 DENA (2014). *Integration der erneuerbaren Energien in den deutsch-europäischen Strommarkt.*
39 Ibid.
40 Ibid.
41 L. Kalinowski and J.-M. Pastor (2013). *L'hydrogène: vecteur de la transition énergetique.* Accessed from www.assemblee-nationale.fr/14/rap-off/i1672.asp
42 Ibid.
43 P. E. Dodds and W. McDowall (2013). The future of the UK gas network. *Energy Policy*, 60, 305–316.
44 Energy UK (2015). *Climate Change Risks & Adaptation Responses for UK Electricity Generation: A Sector Overview 2015.* Accessed January 2016 from www.energy-uk.org.uk/publication.html?task=file.download&id=5466
45 ENEA Consulting, France (2012). *Facts and Figures: Energy Storage.* Accessed April 2015 from www.enea-consulting.com/wp-content/uploads/2015/05/ENEA-Consulting-Energy-Storage.pdf
46 D. Connolly, H. Lund, B. V. Mathiesen and M. Leahy (2010). Modelling the existing Irish energy-system to identify future energy costs and the maximum wind penetration feasible. *Energy*, 35, 2164–2173.
47 Ibid.
48 J. Thijssen (2009). *Natural Gas-Fueled Distributed Generation Solid Oxide Fuel Cell Systems: Projection of Performance and Cost of Electricity.* U.S. Department of Energy.
49 Ibid.
50 Hydrogenics (2014). *Excess Wind Power Turned into Gas in Denmark Using Hydrogenics Technology.* Accessed May 2016 from www.hydrogenics.com/about-the-company/news-updates/2014/02/18/excess-wind-power-turned-into-gas-in-denmark-using-hydrogenics-technology.
51 Department of Trade and Industry (2006). *The Energy Challenge: Energy Review Report.* London: The Stationery Office.
52 Greenpeace UK (2005). *Decentralising Power: An Energy Revolution for the 21st Century* (0–74). Accessed May 2016 from www.greenpeace.org.uk/media/reports/decentralising-power-an-energy-revolution-for-the-21st-century
53 I. Hore-Lacy (2007). Nuclear power. *Nuclear Energy in the 21st Century.* London: World Nuclear University Press. pp. 37–53.
54 T. B. Johannson and J. Goldemberg (2002). A policy agenda to promote energy for sustainable development. *Energy for Sustainable Development*, 6 (4), 67–69.
55 A. J. Bradbrook and J. G. Gardam (2006). Placing access to energy services within a human rights framework. *Human Rights Quarterly*, 28 (2), 389–415.
56 J. Kruger, W. E. Oates and W. A. Pizer (2007). Decentralization in the EU Emissions Trading Scheme and lessons for global policy. *Review of Environmental Economics and Policy*, 1 (1), 112–133.
57 G. Bridge, S. Bouzarovski, M. Bradshaw and N. Eyre (2013). Geographies of energy transition: Space, place and the low-carbon economy. *Energy Policy*, 53, 331–340.
58 World Energy Outlook (2006). *World Energy Outlook.* Accessed February 2016 from www.worldenergyoutlook.org/media/weowebsite/2008-1994/weo2006.pdf
59 International Energy Agency (2006). *Energy Technology Perspectives.* Accessed April 2015 from www.iea.org/publications/freepublications/publication/etp2006.pdf
60 Greenpeace and EREC (2007). *Energy [r]evolution.* Report. 1 January.
61 DOE/EIA (2006). *International Energy Outlook 2006.* Accessed from www.hsdl.org/?view&did=15906
62 World Bank.
63 World Bank.
64 Enerdata.
65 Enerdata.

11 Performance of multi-scale energy systems

Energy planners can better plan energy load forecasts with the optimal performance of technologies installed in buildings. Various renewable technologies are available that produce minimum carbon emissions. One such technology that we will discuss in this chapter is the heat pump.

The chapter provides detail on the working of energy-efficient heat pumps. You will also find out how they can be modelled when determining energy load requirements.

11.1 Modelling technology performance in buildings – the case of heat pumps

Actual load for space and water heaters is determined by a number of factors. The most important factors that determine the load include heat system, occupancy, radiator size, flow rates, thermal mass and building heat loss.

A water heat pump that features under-floor heating with a Carnot efficiency of 55 per cent and source temperature of 5°C has a coefficient of performance (COP) of 4.0, according to the IEA Heat Pump Centre. That being said, when the heat pump is used with a conventional heater, the COP increases to 4.5.[1] The increase occurs because conventional heaters are less effective as compared to under-floor heaters, mainly due to surface area. That's why it is important to consider radiator size when installing heat pumps in buildings.

Compressors in the heat pump are of different types that perform differently. Some of the compressors are high performance, while others are low-performance compressors. These include: reciprocating compressor with oversized heat exchangers, single-speed rotary reciprocating compressor, two-speed compressor with variable indoor fan, and advanced reciprocating or scroll compressor. The variable speed compressor is the most energy-efficient compressor.[2] One example of a variable speed compressor is the Mitsubishi Ecodan air source heat pump.[3]

11.1.1 Coefficient of performance (COP)

An important term that is particularly relevant when it comes to efficiency of heat pumps is the coefficient of performance (COP). Every heat pump has minimum performance requirements reflected in the COP. The values for minimum COP differ for different types of heat pumps manufactured by different manufacturers.

The COP is the ratio between heat energy output (Qout) and energy input (Ein) as a function of time.[4]

$$COP(t) = \frac{\dot{Q}_{out} \ (in \ kW)}{\dot{E}_{in} \ (in \ kW)}$$

Another important term relevant to heat pump energy is the seasonal performance factor (SPF), which represents the COP averaged over a particular season:

$$SPF = Q_{season} / E_{season}$$

Performance of heat pumps depends on cooling demands and seasonal heating, which vary over time. Moreover, there are a number of other factors that determine the performance, including heat pump power rating, temperature lift over time, time-control strategy of the heat pump and operating mode.[5] Moreover, system boundary characteristics also directly influence performance by influencing the load.

Electrical input is generally metered over five-minute time steps, after which heat output is measured using a flow and temperature rate that is connected to a heating system.

11.1.2 *Heat pump models*

Heat pump models are used to model the dynamics of heat pumps for optimising performance output. The following are the three most common types of heat pump models that are discussed in the literature:

a) *Equation fit model* – this model uses a statistical curve fit approach to data modelling. There is no physical modelling of the components.
b) *Deterministic model* – the model uses thermodynamic theory as well as heat mass balance equations to model the data.
c) *Parameter estimation model* – the model technique consists of using simplified mass heat balance equations. However, the estimates of the model parameter are calculated using parameter optimising techniques and the manufacturer's data.

Remember that the models try to capture the dynamics of heat pump performance – but the real performance of the heat pump may not always match. Gaps have been reported about the theoretical construct of the model and the actual performance. In order to understand the variances, we have to get an understanding of how heat pumps work in different conditions.

Different building simulation software is available that can be used for modelling such as Energy Plus, TRANSYS, etc. The simulation software differs when it comes to degree of accuracy. For instance, Energy Plus can accommodate a large number of variables and air/heat flow algorithms. The software can also incorporate a wide number of variable inputs such as material properties, building dimension and occupancy schedules, to name just a few.

Here we will apply a simple theoretical model that has been presented by different scholars to simulate an air source heat pump installed in a building system.[6,7,8] The model has been used in Simulink and Matlab in different past studies.[9,10] It has been calibrated with available datasets relating to heat pump trials.

Because functionality of the models can be validated by comparing the results with real time data, a sensitivity analysis has been performed with varying parameters to determine the impact of different constraints on performance. The information gathered in this way can be used to make practical decisions.

Figure 11.1 depicts the physical components, boundaries and variables that are incorporated into a heat pump model and the building zone model.

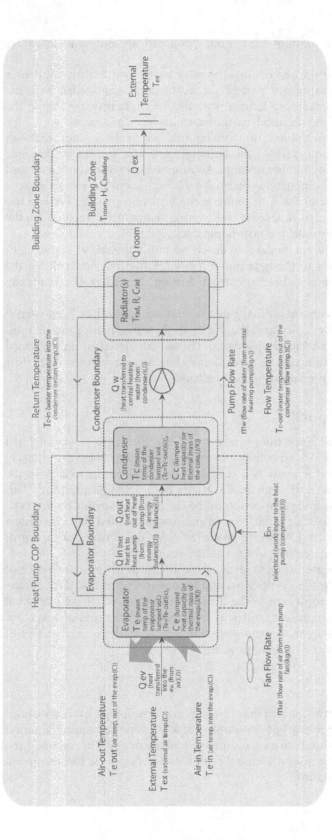

Figure 11.1 Heat pump and building zone model diagram.

General equations and notations have been proposed in different literature studies.[11,12,13]

A heat pump module is comprised of three primary parts: evaporator, condenser and heat pump cycle. The evaporator part of the heat pump is modelled with the assumption of uniform temperature in lumped volume. The heat transfer rate of the evaporator is represented by the thermodynamic heat difference equation.

The condenser is lumped in volume in a similar manner to the evaporator with a water side present in the boundary of the building zone and the refrigerant side located inside the boundary of the heat pump. Assumptions for the condenser and evaporator are the same, i.e. homogenous thermal mass with constant temperature.

The third main component of the heat pump is the heat pump cycle. We can use the Practical Carnot equation that is based on the average temperature recorded in the condenser and evaporator to model the heat pump cycle.

The next stage of the heat pump model is to develop a building zone with the intention of connecting a heat load to the heat pump model. A thermal network for the building or room heating zone has been constructed using the orders of complexity outlined by Rabl (1988).[14]

The Matlab-Simulink solver for ordinary differential equations has been used.[15] The Runge-Kutta Ode45 solver calculates the x state at the subsequent time-step utilising the current state of x and its derivative x. This is calculated using a defined time-step size dt.[16] The program can be used to calculate air-out temperature, flow temperature, return temperature and room temperature.

11.2 The concept of load profiles

Load profiling has been used since the foundation of the Electricity Council Load Research Programme in the 1950s to collect and analyse data. That being said, using load profiles for electricity settlements is a relatively new concept especially in the UK. The concept was introduced through the Electricity Pool '1998 Programme'. This programme was set up to increase competition in the electricity distribution market.

To avoid large costs due to Half Hourly metering, it was decided that consumers with a maximum demand of 100kW would be settled using the concept of load profiles by looking at the existing meter readings of the customer.

The load profile can indicate different data types. It can point to a reference of derived data forms such as Profile Coefficients and Regression or to raw data depicting customer consumption and demand. The different types of data have one thing in common, in that they all represent electricity consumption patterns of a specific market segment. The patterns or shapes can show half hourly, daily or yearly usage of data that differs for each category of load profile.

11.3 Peak load factor

Peak load factor is an important concept that can help in determining electricity usage. It is defined as the ratio (in per cent) of the number of kWh supplied during a period to the number of kWh that would have been supplied had the maximum demand been maintained throughout that period. An annual peak load factor can be represented in the form of the following equation.

$$LF = \frac{\text{Annual consumption (kWh)}}{\underset{\text{demand (kW)}}{\text{Maximum}} \times \underset{\text{in the year}}{\text{Number of hours}}} \times 100$$

11.4 Modelling energy demand and supply at the temporal–spatial scale at different levels in the energy system

Energy demand and supply can be modelled within an energy system at different levels of temporal–spatial scale. These include distribution network, operator, local electricity substation and grid supply point demand.

Development and growth of urban systems has remained somewhat constant for about six millennia. Today, however, individual cities have reached proportions that were unimaginable about 50 years ago. Urban towns and cities have grown about a thousand times during this time period. They are multi-scalar across time and space, with such a great extent of interaction between top down and bottom up that has never been seen before.

Managing resources in urban areas has become a burning topic among politicians, scientists and decision makers related to different economic segments, as well as the general public. Although research is conducted in transdisciplinary teams, most of the existing research remains essentially sectorial with a focus on water, pollution, energy, metabolism and the like.

Differential equation models could also be used that reflect 'top-down' parts of behaviour including institutions, rules, external conditions and other similar conditions. A multiscale spatial structure could also be incorporated as a set of GIS layers featuring transition matrices that establish the conditions for spatial structure changes.

A lot has been written shedding light on how to model energy demand for residential, industrial and commercial buildings. The existing literature has shed light on this matter through distinct categorisation and development of energy supply models for predicting the evolution of renewable energy.

Additionally, a lot of modelling methods simulate the production, distribution and consumption of energy. That being said, all the different models have some strengths and weaknesses in terms of practical applicability. In order to overcome the weaknesses, there has been a recent development in coming up with integrated models. A big challenge in introducing such a holistic model is how to capture different spatial scales and simulate the dynamics at micro and macro levels.

The advantage of the DEAM model[17] is that it can be applied to different cities, countries and regions worldwide. Moreover, the model can be used in various research and commercial projects.

For instance, the model found application in determining energy flows for agents of a local substation in order to predict energy loads using a set of data from Western Power Distribution (WPD) and Distribution Network Operators (DNOs). In addition, the model has also been used to examine the point demand of grid supply.

To sum up, DEAM methodology consists of the following main points:

- disaggregation of net energy demand of a particular area into various end-use categories;
- application of suitable load profiles for energy allocation to consumers to day of week, month of year and half hour of day. The best thing about DEAM is that it can take into account any number of end users and load profiles;

- simulation of multiple scenarios after incorporating economic, technical and social development;
- energy representation of demand and supply in the present and future energy system scenarios at a half hourly temporal resolution and multiple spatial scales;
- investigating the consequences of energy system changes at a disaggregated temporal resolution of half hourly temporal resolution.

DEAM calculates end user demand by using a hybrid approach, i.e. combining two distinct methods: top down, such as non-space heat demands, and bottom up, which includes space heating/cooling demands. The top-down approach analyses demand values by end users as well as by time, using activity profiles that describe the utilisation of energy over time by different end users. The bottom-up approach, on the other hand, builds physics-based methods that have the ability to introduce new demand types.

11.4.1 Energy demand calculations and databases

a) Non-space heat demand

The DEAM model mentioned in the previous section can be used to determine non-space heat demand by the end user. The approach requires annual energy demand value for every end user. The data can be collected from published energy statistics or other online sources. Demand can be calculated for different end users as well as for different fuels.

The annual value that is used for calculating non-space heat demand must be disaggregated temporarily by first dividing the average daily energy demand by number of days in a year and then by half hourly values. This is achieved by using activity profiles that are specific to the end users.

The equation given below can be used to determine end-use energy demand or load profile for domestic lighting.

$$\text{End User Demand} = \text{Daily Mean Demand} \times \text{AP}$$

In the above equation that is used to calculate load profile, the Daily Mean Demand is the average daily demand for each end user, while AP is the normalised activity profile that incorporates the hourly, weekly and seasonal changes in demand.

Another equation that can calculate the normalised value of energy using the DEAM approach is given below.

$$\text{Normalised Value of Energy} = \text{Sector month} \times \text{End use month} \times \text{Sector day} \times \text{End use hour}$$

Activity profiles that are developed using the above equation contain values that depict the relative amount of energy utilised over different points of time in a day, week and year that are multiplied to offer a normalised value for every half hourly time step. The method can help in determining the variation in energy use over multiple temporal resolutions.

Figure 11.2 shows an example of a daily activity profile for each sector – domestic, industrial, service and transport – for January (month 1) and for Tuesday (weekday 2).

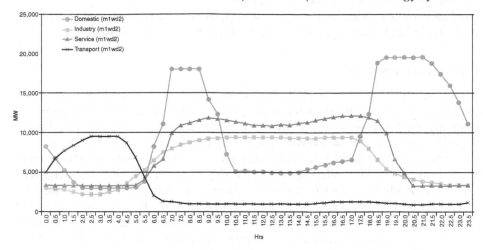

Figure 11.2 Example of daily activity profile for a weekday in January.[18]

The equations that use the DEAM approach make use of the same data but produce different outcomes. An advantage of using the approach is that bespoke equations can be developed that reflect different scenarios. Activity profiles that are determined using the equations can then be used to develop load curves.

b) Space heat demand

Space heat demand can be calculated through a bottom-up DEAM approach that makes use of cooling and heating demand. This is different from the demands calculated using the top-down technique in that they are influenced by weather in addition to building characteristics.

The model determines the energy required to heat the building or part of a building to a specific temperature. It uses a heat loss coefficient that is normalised to the building floor area ($W/K/m^2$) while the incidental gains are deducted from the heat demand.

The assumption of the model is that each person consumes a specific wattage of electricity, some of which is wasted, such as from lighting, cooking and solar panels. This requires calculation of heat loss per person per hour, glazed area per floor area for solar gains, and the proportion of power wasted as heat for each end use, as well as the amount of this heat that is useful for heating a dwelling. The wasted heat from every end user necessitates inputs from the top-down section of the model.

c) Non-domestic agents

Non-domestic agents are of two types that include:

1   *services agents* – e.g. offices, shops, health, warehouses, education, leisure and hotels;
2   *industry agents* – e.g. principal subsectors such as iron and steel, minerals products, non-ferrous metals, chemicals, electrical and instrument engineering,

mechanical engineering and metal products, drink and tobacco, vehicles, food, leather, clothing, textiles, paper, publishing, printing, construction, other.

DATABASE PREPARATION PROCESS

At the moment there are no available comprehensive databases for non-domestic agents. This is partly due to the heterogeneity factor mentioned above that increases the costs incurred in collecting the data and partly due to commercial sensitivities' lower political priority due to less energy usage by the non-domestic sector.

Due to the complexity involved in gathering data about non-domestic agents, different data sources are utilised by collecting and processing to predict building characteristics and energy demands. This data forms the foundation from which a more meaningful database is created that provides the required information about energy demands and consumption apropos heat, electricity and gas. In other words, database preparation for non-domestic agents requires coupling, given primary data with default assumptions, as shown in http://uksim.info/uksim2015/data/8713a195.pdf.

Figure 11.3 depicts some of the databases and relations that are used for modelling a substation. The industrial partner section of the project offers a connectivity database that is used as a master list, consisting of connected agents and other data sources. These are then matched using different techniques such as agent and postcode name searching.

An example of a database that is appropriate for the non-domestic sector includes the Display Energy Certificate online database (Display Energy Certificate Database).[19] This database provides information of different types of public building and associated annual electricity demand. Another comprehensive database is the UK Marketing Database (UKMD) that consists of data from about 4.2 million enterprises describing the nature of their work, including building characteristics such as size, energy demand profiles and the number of employees in the building.[20]

Now let us look at the actual steps that are involved in the database preparation:[21]

S1 define all required inputs and the structure of model inputs;
S2 obtain and analyse extracts from all primary data sources;
S3 prepare source to target mapping, highlighting any gaps and issues;
S4 create and maintain the data dictionary;
S5 develop estimating rules and assumptions to plug any gaps.

For more information on the process please see http://uksim.info/uksim2015/data/8713a195.pdf

For information on the DEAM model please see www.ucl.ac.uk/energy-models/models/deam

For information on SpDEAM please see Ed Sharp, PhD, 'Spatiotemporal disaggregation of GB scenarios depicting increased wind capacity and electrified heat demand in dwellings'. The DEAM model has been used as the base of the SpDEAM model.

11.4.2 Modelling demand response

The appliances that can potentially help in reducing energy demands in the future grid without compromising the comfort and convenience of the occupants, and reduce energy consumption, include the washing machine (WM), dishwasher (DW) and dryer (DY).

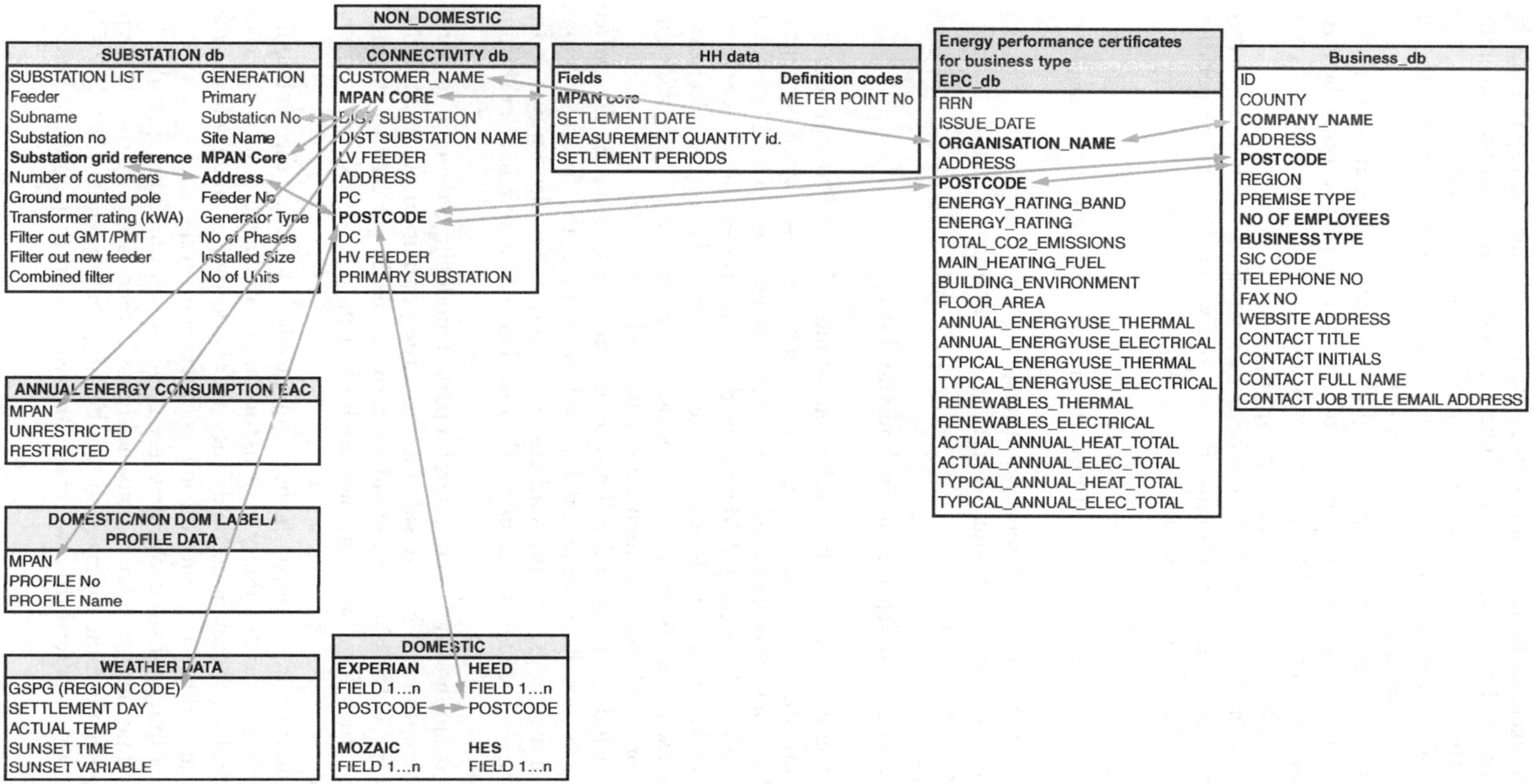

Figure 11.3 Logic diagram for linking databases.[22]

The aim of the exercise that follows is to examine load shifting capabilities by scheduling the connected appliances of a dwelling group so as to adjust the consumption load curve. The objective load curve is established by the load demand in a particular period of time, after which an algorithm is then used to minimise the gap between the estimated and the actual consumption loads.

During different simulation time steps, the appliance start events are determined stochastically. Power demand is added to the net total for the dwelling group whenever an appliance is used. Demand profiles differ significantly in a day between different groups and between different days for the same group.

Matrix Laboratory's MatLab vR2010b has been used, while the appliances selected are operated as per consumer demand only. Energy and water during the operation are calculated using the following factors:

- frequency of operation;
- programme and temperature selected in combination with type and amount of detergent;
- efficiency of machine under different conditions;
- time span in lower power mode such as start delay + standby;
- additional rinse option selected;
- load size used;
- operation frequency that depends on the household size.

The results of EuP-Studies of the washing behaviour of more than 2,500 individuals in ten countries portrays the finding that the 40° C programmes are used by 37 per cent of all respondents.[23] The next two most used temperatures are 60° C (used by 23 per cent) and 90° C (used by about 7 per cent). Additionally, the study also examined whether or not a machine has a low power mode function, i.e. start time delay or pre-select function, and how it is used. This information can be used to calculate activity profiles that will then be used for calculating load demand in a particular building.

Model inputs include a group of dwellings connected to a system, the diversified load representing expected consumption for each appliance, the forecasted load for this number of dwellings, and the load curve demand that is also known as the 'objective' load curve. The model determines the final load of the system after shifting the appliances.

Apart from model inputs, the other important part of modelling a demand response is defining the assumptions. When considering the entire population, we assume that 75 per cent of the individuals in a household are active while 25 per cent are inactive. Also, all buildings stock for a weekday, and each of them have a washing machine, dishwasher and dryer.

Once the assumptions are defined, the next thing to do is to define the availability of individuals as the a priori possibility in utilisation of an appliance. This is based on the activity data profiles of each household. They are also known as unitary load cycle or the power demand of the appliance during its functioning cycle.

We developed a unitary load cycle for every appliance in functioning modes. At first, we assumed that, in a simulated household stock, there are only five different types of washing-machine functioning modes or heating phases, namely 30° C, 40° C, 50° C, 60° C and 90° C. Afterwards, we made an assumption that only the heating phase is changed by the temperature selection. We then defined a corresponding unitary load cycle for each functioning mode and then calculated the forecasted load curve through the load curve that had accrued.

As a result, the model is provided with four objective load curves. For each of the load curves, the algorithm aims to close the gap between the final load curve and the objective curve through controlling and modifying the use of controllable appliances such as a washing machine.

Results of the modelling demand procedure

We modelled the predicted load curve of consumption for X dwellings, each of which had the three appliances WM, DY and DW. The basic model is comprised of different objective load curves, as presented in Table 11.1.

Our findings suggest that as the number of households increases, the final load for the system becomes smoother and similar to that of the objective load. In Table 11.1 you can see the diversified load that is caused by washing machines.

Forecasted loads is the total net load expected in the system, in our case the sum of the electricity and the washing machine's load. Figure 11.4 shows the final load curve formulated based on the desired objective load curve for all three cases considered above. The objective functions used in the first two cases raised the load in the early morning hours. This conducted a uniform distribution of the consumption compared to the situation in the third case.

In this modelling exercise, a hybrid optimisation method called HGPD (consisting of three optimisation algorithms: genetic, particle swarm and downhill) was used. In terms of accuracy, convergence rate, stability and robustness, it was proven that the hybrid method could provide slightly better accuracy, depending on the optimisation problem. The HGPD algorithm is a type of stochastic optimisation algorithm that can search a complex and uncertain area. This exercise in modelling demand response can assist in determining the future implications on smart grids of the demand-side response.

Table 11.1 The objective load curves for different cases

Case no.	Objective load curve
Case 1	
Case 2	
Case 3	

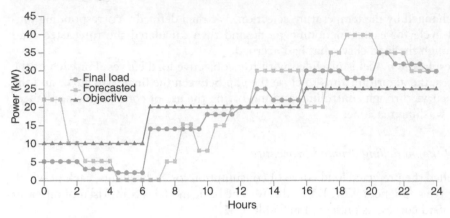

Figure 11.4 Final load curve forecast for all cases.

11.5 Interrelation with other resources: energy–water–land nexus

The energy–water–land nexus must be accessed from an energy perspective. The needs of land and water can be expressed as withdrawal and consumption for electricity generation that can be estimated by climate and region scenarios.

Water use can be separated into withdrawal and consumption functions. Note that water here refers to freshwater. Use of saline water from the sea is not considered during demand modelling despite the fact that saline water, unlike freshwater, is available in abundant quantities.

a) Representation of space and time

Modelling also makes it possible to subdivide the study area into sub-areas. This is made possible by using a pyramid structure. In this way a whole country in different regions or municipalities can be represented. Results can either be examined at a micro level for small units or at a more macro level for higher level units. The latter consists of all results of its subordinated units.

Entities can be modelled as knots or an areal extent that is only recorded as figures. The exchange between the entities is not provided, except for electricity, which is represented as an integrated system for different study areas.

b) Electricity for water

Water-related services include the water supply that is used for industrial, agricultural and domestic purposes. Moreover, it can also relate to the supply of water for treatment of municipal wastewater.

Each supply need for the different groups, as well as the amount of wastewater that must be treated, are calculated for every time-step and sub-area having a specific growth factor using the following equations:

$$WC_{c,a,i+1} = WC_{c,i} (1 + GRwc_{c,a})$$

$$WW_{c,a,i+1} = WW_{c,i} (1 + GRww_{c,a})$$

where

$WC_{c,a,i+1}$ = Water consumption of category c in area a, at time step i + 1 (hm^3)

$GRwc_{c,a}$ = Growth rate water consumption of category c in area a (hm^3)

$WW_{c,a,i+1}$ = Water withdrawal of category c in area a, at time step i + 1 (hm^3)

$GRww_{c,a}$ = Growth rate water withdrawal of category c in area a (hm^3)

Energy demand to pump cooling water for power stations is not considered in the above equations since it is already incorporated into the overall efficiencies of the stations.

Calculating net power supply requirement

Electricity demand is comprised of a base demand that includes industrial, commercial, residential and public demand, and a demand for water-related services.

Electricity demand can be calculated for the whole study area while making the assumption of unlimited connection between separate sub-areas.

PC_i = Total power consumption at time step i (e.g. GWh)

PCb_i = Base power consumption at time step i

PCw_i = Electricity demand for water services at time step i

Total power supply requirement is calculated based on the total electricity demand and the electricity grid losses. The loss factor can remain constant or vary over a period of time on account of efficiency changes in the transmission grid.

$$TPSR_i = \frac{PC_i}{1 - Len_i}$$

where

$TPSR_i$ = Total power supply requirement at time step i
Len_i = Electricity grid losses at time step i

Determining power supply capacity

Power supply capacity can be subdivided into five different categories, namely hydro, photovoltaic, wind, nuclear and thermoelectric power generation. The thermoelectric group consists of four subcategories: coal, oil, natural gas and biomass.

Capacity factors are defined by users of the model. These factors function as the representation of long-term climatic changes and seasonal variations. Capacity factors need to be defined for each technology type, time step and sub-area. The capacity factors of nuclear, hydro, photovoltaic and wind generation are supposed to be fixed and serve as an input for determining the power output.

Thermoelectric capacity factors remain flexible when calculating the power supply. These include: photovoltaic, wind, nuclear, hydro and thermoelectric. Thermoelectric capacity is used to match the demand with the remaining power supply. This requires defining just the feasible maximum thermoelectric capacity factor.

The following equation can be used to calculate actual capacity factors. The first step of the calculation involves calculating the factor of supply from photovoltaic, nuclear and wind generation which is calculated based on their capacity factors:

$$PS_{t,a,i} = GC_{t,a,i} \times Fc_{t,a,i} \times Ft$$

where

$PS_{t,a,i}$ = Power supply by technology t, sub-area a and time step i (GWh)

$GC_{t,a,i}$ = Generation capacity of technology t, sub-area a and time step i (GWh)

$Fc_{t,a,i}$ = Capacity factor of technology t, sub-area a and time step i (GWh)

Ft = Time factor

The main assumption in the above equation is that all the generated electricity can be supplied to the grid while taking into account that fluctuations in supply of energy generated through photovoltaic and wind generation is subject to fluctuations. Also the equations are only applicable if the proportion of energy supplied by photovoltaic and wind generation is relatively small.

The second step involves calculating the remaining supply requirement.

$$RPSR_{1,i} = TPSR_i - \Sigma_a (PSpv_{a,i} + PSw_{a,i} + PSn_{a,i})$$

where

$RPSR_{1,i}$ = Remaining power supply requirement at time step i (GWh)

$PSpv_{a,i}$ = Power supply photovoltaic of sub-area a and time step i (GWh)

$PSw_{a,i}$ = Power supply wind of sub-area a and time step i (GWh)

$PSn_{a,i}$ = Power supply nuclear of sub-area a and time step i (GWh)

In case the hydropower capacity is not sufficient to supply the remaining supply requirement, the following equation can be used to calculate the power supply from hydropower. Note that the remaining power supply requirement (RPSR) is updated by subtracting the power supply from hydropower.

$$RPSR_{2,i} = RPSR_{1,i} - \Sigma_a PSh_{a,i}$$

where

$RPSR_{2,i}$ = Remaining power supply requirement at time step i (GWh)

$PSh_{a,i}$ = Power supply hydro of sub-area a and time step i (GWh)

The last step in calculating the power supply requirement is to test whether the thermoelectric capacity is sufficient for the remaining requirements and is subject to the maximum capacity factor. In such a situation, thermoelectric generation is distributed evenly between the sub-areas that are subject to the installed capacity.

$$PSt_{a,i} = \frac{RPSR_{2,i}}{\Sigma_a GCt_{a,i}} GCt_{a,i}$$

where

$PSt_{a,i}$ = Power supply thermoelectric of sub-area a and time step i (GWh)

$GCt_{a,i}$ = Generation capacity thermoelectric of sub-area a and time step i (GWh)

The thermoelectric capacity factors are calculated to monitor the utilisation

$$CFt_{a,i} = \frac{PSt_{a,i}}{GCt_{a,i} Ft}$$

where

$CFt_{a,i}$ = Capacity factor thermoelectric of sub-area a and time step i

Ft = Time factor (e.g. 720 hr/month)

In the event that the hydropower capacity is enough to supply the remaining power supply requirement (RPSR), then no additional thermoelectric power generation is required. In this case, the procedure undertakes hydropower generation instead of thermoelectric power generation.

c) Water for electricity

Water consumption and withdrawal for electricity generation are recorded separately. Every power generation category that includes the subcategories coal, biomass, natural gas and oil is allotted a consumption and withdrawal factor for each unit of electricity generated.

The water use factor consists of the production and extraction of fuel as well as the onsite generation of electricity. Water use for each sub-area and time step is determined based on the power supply requirement by the specific technology.

$$Wc_{t,a,i} = PS_{t,a,i} \times Fwc_t \times Fc \quad WW_{t,a,i} = PS_{t,a,i} \times Fww_t \times Fc$$

where

$Wc_{t,a,i}$ = Consumption of technology t, in sub-area a, at time step i (m3)

$PS_{t,a,i}$ = Supply by technology t, in sub-area a, at time step i (GWh)

Fwc_t = Technology-specific water consumption factor (m3/MWh)

Fc = Conversion factor

$WW_{t,a,i}$ = Water withdrawal of technology t, in sub-area a, at time step i (m3)

Fww_t = Technology-specific water withdrawal factor (m3/MWh)

d) Land for power generation

Land utilisation for electricity is determined based on a constant factor for each technology and power supply requirement for specific technology.

$$LU_{t,a,i} = PS_{t,a,i} \times Flu_t \times Fc$$

where

$LU_{t,a,i}$ = Land use of technology t, in sub-area a, at time step i (m2)

$PS_{t,a,i}$ = Power supply by technology t, in sub-area a, at time step i (GWh)

Flu_t = Technology-specific land use factor (m2/GWh)

Fc = Conversion factor

Total land use in each of the sub-areas is estimated for each time step so that the land use for power generation is correlated to the total land use. Percentage growth rates are used as a projection base to calculate future land use.

e) Limitations of the model

The model that is used to examine the impact of power generation on land, water and other resources has certain limitations. First, the model is used only to address a part of the whole water–energy–land nexus. For instance, energy use for land is not taken into account, which in the case of farming is mainly based on fuels not considered in the model, unless they are used for electricity generation.

Second, the model neglects the relationship between land and water use since this would require a real assignment of land use, as well as examining the effects on water balance that is beyond the scope of the model.

The third limitation of the model is that it uses average capacity factors for different sub-areas, and fluctuations in the availability of single power stations cancel each other out. The issue will have significant effects if the whole generation category such as thermo-electric generation is zero or close to zero. This shows a severe disparity between demand for electricity and the available capacity, which is unlikely to happen in a real case scenario.

For more information on the modelling and its application see the reference http://uksim.info/ems2015/data/0206a266.pdf, which provides an analysis of a number of modelling scenarios for the water–energy–land nexus for Brazil.

11.6 Regional-level/super grid energy systems

A super grid can be defined in general terms as a method of connecting production zones that have a high potential of renewable energy generation with high demand zones.

Another definition of a super grid is:

> SuperGrid relies on the mechanism of cross-system electricity exchange (export and import) across systems with different intermittency sources, balancing technologies, and demand patterns. This mechanism makes it theoretically possible to handle large-scale penetration of intermittent resources without any short to medium-term need for storage or demand flexibility.[24]

An example of a super grid is the North Sea region, where there is great potential for the generation and export of wind electricity to high electricity demand areas in central European countries.

Salient characteristics of a super grid system

A number of studies have shed light on the characteristics of a super grid system.[25,26,27,28] Some of these include:

- high priority for the construction of electricity corridors or electricity highways for prioritised corridors;
- super grid reliance on direct current (DC) cables instead of the traditional alternating current (AC) cables;
- greater integration of different production and consumption centres of renewable energy especially across Europe and North Africa;
- potential to produce excess electricity that can be exported;
- replacement of the need for local production as individual countries can satisfy their electricity needs with imports;
- allowing cross-border distribution of electricity.

The main benefit of the super grid is optimal resource utilisation due to the cross-border trade in electricity. Another key benefit of the super grid is that it allows greater integration of renewable energy sources as it contributes to a reduction of the carbon footprint at an international level.

There are a number of experts worldwide who analyse super grids. One of them is Dr Gregor Czisch. His pioneering work on European super grids has had a worldwide impact. His dissertation on powering Europe with renewable energy has been published by the Institution of Engineering and Technology as a book entitled *Scenarios for a Future Electricity Supply*. It is available as a free download from Kassel University (www.agenda21treffpunkt.de/doku/sign.php?sg=Czisch-2006).

There are a number of projects as well that are investigating different characteristics.

One of these is the TradeWind project, which investigated the impact of increasing wind capacities on the flow of electricity between countries and found that upgrades to specific interconnections between countries in a timely manner would reduce total system costs.

A number of case studies are provided in the following section. Key messages from the current research include the following:

- transmitting the generated electricity over long distances to load centres with the use of high voltage direct current (HVDC) transmission lines can help achieve energy supply security by generating electricity from suitable sites where renewable sources of energy are abundant to where it is needed;
- it can help to facilitate the integration of renewable energy on a large scale;
- it can help develop the political, legal and regulatory conditions for a unified transnational and intercontinental market with the aim of providing reliable and affordable electricity and economic development.

11.7 Case studies

China–East Asia super grid

A super grid is an indispensable tool that is used to make renewable energy, and to share vital resources such as conventional generation, dispatchable hydropower, pumped storage, peaking turbines and new energy storage methods such as vehicle-to-grid and spinning reserve. A super grid enables a wide area to share resources, reducing the total generation capacity that is needed. An Asian super grid can serve as the core aspect of the overall Asian shift to a sane and sustainable energy and climate future.[29] Figure 11.5 shows an Asian super grid that ties together countries that have difficult relationships and will require political action. Such a regional super grid would eventually be connected to Europe and Africa. An Asian super grid can evolve from existing and under-construction HVDC transmission lines that can serve part of China by interconnecting these power lines into a limited but highly useful super grid, which can then extend to Northern China and Southeast Asia.

Concept: in order to give way to economic efficiency, the existing HVDC infrastructure should become a part of the Asian super grid. In order to see how feasible these actions are, there is a need to do simulations and scenario analysis, for which one requires the right kind of tools. There is a growing need to do scenario analysis in order to identify how the injection/removal of power at the AC grid node points will aid system stability. Also, there is a need to utilise advanced methods for optimising renewable generator locations through comprehensive weather data analysis. Furthermore, system simulation modelling is also required that includes integration and optimisation of local generation and two-way energy flows at the level of residential units, industrial facilities or community micro grids. This basically consists of:

- rooftop solar PV;
- cogeneration and other biomass fuel-drive local generation (perhaps a tax on waste heat would favour cogeneration).

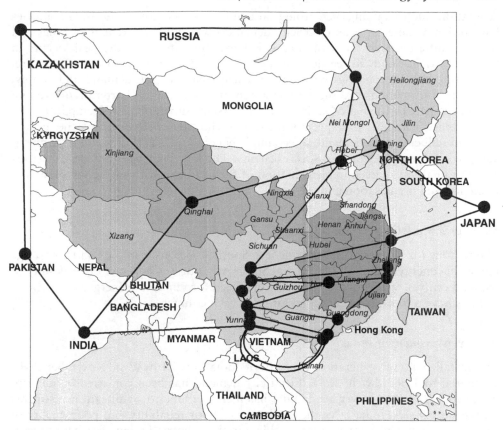

Figure 11.5 Asian super grid.[30]

New, small dispatchable hydro (growing rapidly in China) needs a regulatory framework. The question is whether it is worth making these generators partially dispatchable. More analysis is needed.

Gobitec and Northeast Asia super grid in the Gobi Desert[31]

The Gobitec project is similar to the DESERTEC project. It aims to deploy renewable energy plants in the Gobi Desert to meet electricity demand. Participating countries are Russia, South Korea, China, Japan and Mongolia.

It has been concluded that in order to ensure the effective flow of electricity generated in the Gobitec project, cross-border transmission lines should connect the Gobi Desert with Irkutsk in the north, Shanghai and Seoul in the south and Tokyo in the east of the Asian Super Grid (ASG) region. The high potential for renewable energy technologies in the Gobi Desert of Mongolia and large amounts of hydropower potential in Russia determine that these two countries should export renewable energy, while all other countries will import energy from the Gobitec project.

All participating countries benefit from the interconnection. One benefit is balancing supply and demand because of spatial diversification. This can help the electricity generation due to varying meteorological conditions. Moreover, the ASG countries have different times of peak electricity demand because of different time zones and seasonal differences in the load curves. For example, in Japan the peak occurs in the summer. In South Korea demand peaks occur during winter times. Therefore, connecting several countries with one grid helps balance the demand and supply. For more information read the following report: www.energycharter.org/fileadmin/DocumentsMedia/Thematic/Gobitec_and_the_Asian_Supergrid_2014_en.pdf.

There are a number of expectations, from technical benefits such as energy savings and reducing peak demand, to economic benefits such as short-term investment, along with environmental and social benefits. The key challenges remain consensus and sustainability, agreed implementation of the roadmap with the action plan, and implementing organisation and rules.

Super grid modelling for South Asia

The reasons for a super grid deployment in South Asia consist in the fact that a large renewable energy potential is located in South Asia but it is not in close proximity to the load centres. In addition there is a strong need for renewable energy.

Let's have a look at the energy situation in each country.

Energy assessment of India

The installed power generation capacity of India was 210GW in November 2012 compared to 1,362MW in 1947. Electricity demand has been consistently outstripping supply, even as power availability has increased due to significant investments on the supply side. India faces the challenges of poor reliability and poor quality of electricity, leading to frequent load shedding in the country. Indian electricity generation mostly comes from nuclear, hydropower and conventional thermal.

The predominant source of electricity generation is coal, which recently showed a significant shortage. Coal results in high-content carbon emissions, and the refining and extraction of the resource is not going on at a level that could cope with the coal-utilisation factor of the generating stations. The main renewable energy sources in India are wind energy, solar energy, biomass and waste, and hydro energy. Importing gas is not a feasible option, as renewable energy resources are located far from the load centres.

Energy assessment of Pakistan

The power situation in Pakistan is one of the most severely criticised services rendered by the government of Pakistan, due to which load shedding and power outages are common. Electricity generation, transmission and distribution run under the auspices of two vertically integrated public sector utilities (PSUs): the Water and Power Development Authority (WAPDA) and K-Electric. Around 20 independent power-producing companies contribute significantly in the field of electricity generation in Pakistan.

The Pakistan grid can be directly connected to the northern and western grids of South Asia via HVDC links, leading to a more dynamic and extensive grid. Electric power can be easily transferred across the border via a DC link, thereby making the Pakistan national grid strong and resilient and a reliable power supply to its population. Electricity generation comes from fossil fuels, nuclear, hydropower and renewables. Moreover, Pakistan's supply side has seen no growth over the years, with demand increasing consistently, which is a growing concern.

Energy assessment of Bhutan

Bhutan is a relatively small country lying in the eastern Himalayas, with Thimphu as its national capital. Bhutan is landlocked between China in the North and India in the other three directions. As of 2012, only about 70 per cent of households had access to electricity, and hydro power potential is around 30GW. Bhutan currently exports more than 75 per cent of its generated hydroelectricity to India.

Energy assessment of Bangladesh

Bangladesh is located at the apex of the Bay of Bengal, bordered by India and Burma (Myanmar) on its north, west and east, and separated from Nepal and Bhutan by India's narrow Siliguri corridor. Bangladesh depends heavily on the availability of gas for its generation of electricity.

Energy assessment of Sri Lanka

Sri Lanka is an island country in the northern Indian ocean off the southern coast of the Indian subcontinent in South Asia. Electricity needs are met mostly by generating electricity using petroleum-based thermal sources and hydro-based generation.

Energy assessment of Nepal

Biomass and hydropower are two indigenous and highly potential energy sources in Nepal. Fuel wood supplies nearly 80 per cent of the cumulative energy demand, but the country has a potential for producing nearly 83,000MW of hydroelectricity, out of which only about 280MW has been developed. Nepal does not utilise this electricity potential; it supplies only about 10 per cent of total energy consumption.

A super grid in this region could improve the flow of electric power between nations and provide the capability to interconnect the asynchronous transmission systems of the subcontinent.

Super grids (MENA countries)

The production of solar thermal power is based on a parabolic through power plants. In other words, the plants are able to monitor the diurnal course of the sun through a single axis system. Moreover, they are also able to store up to seven hours of production at normal capacity. What's more, electricity generated through a solar thermal power plant is influenced by atmospheric conditions and irradiance.

Solar thermal makes use of direct sunlight. MENA regions such as steppes, deserts and savannas receive a maximum amount of sunlight. However, a problem is that fumes, humidity and dust may block sunbeams from reaching the solar panels.

The desert areas of the MENA region, northern parts of China and South Western United States have high values of Direct Nominal Irradiation.[32] At this point, it must be noted that trading in solar electricity from the MENA region to Europe or even to China and the US is costly. This is because conversion stations and HVDC cables will be needed to transfer the energy over a long distance. For this reason, a high and stable demand should be present before setting up a super grid in a particular location. This is not a modelling issue but a problem that is economic in nature.

Super grids (Europe)

European IEN is at present the largest interconnected grid in the world. The system consists of 34 countries and has developed gradually over time from independent grids that served local regional areas. The history of grid integration in Europe started in 1952 after the foundation of the Union for the Coordination of Production and Transmission of Electricity (UCPTE). This union was formed between eight European nations comprising Austria, Belgium, France, Germany, Italy, Luxembourg, the Netherlands and Switzerland.

The main aim of UCPTE was to connect the electricity networks of the respective countries through efficient utilisation of limited energy resources which would also help the countries in the event of an emergency.[33]

By the year 1987 about 16 countries were connected through their super grid. During the Yugoslav Wars in 1991 the grid was subdivided into two zones that were later reconnected in 2002. After the trade liberalisation of the power generation market in 1999, UCPTE was renamed as the Union for the Coordination of the Transmission of Electricity (UCTE) to focus on the distribution aspect of the power grid. The objective of the new organisation was to ensure enhanced coordination between Transmission System Operators (TSOs) and also to plan for the future extension of the grid.[34]

UCTE combined with other TSO associations in Europe in 2009 that included the Association of the Transmission System Operators of Ireland (ATSOI), Baltic Transmission System Operators (BALTSO), UK Transmission System Operators Association (UKTSOA) and Nordic Electricity (NORDEL). The resulting institution was named as the European Network of Transmission System Operators for Electricity (ENTSO-E).

The functions of ENTSO-E can be categorised into three parts: to ensure the secure and reliable operation of the connected electricity grids that supply power in Europe, development of a functional IEM, and the maximum integration of electricity obtained from renewable sources.[35] ENTSO-E consists of 42 TSOs from 34 countries that interconnect five synchronous areas and distribute electricity to 532 million people in Europe. Every synchronous area comprises countries that are connected through high voltage transmission lines and synchronised to similar operating frequencies. Countries that comprise different synchronous areas are given below:

- *Baltic synchronous area*: Estonia, Latvia, Lithuania;
- *Nordic synchronous area*: Denmark (east), Finland, Norway and Sweden;
- *Irish synchronous area*: Ireland;
- *UK synchronous area*: Great Britain;

- *Continental Europe synchronous area*: Austria, Belgium, Bosnia-Herzegovina, Bulgaria, Croatia, Czech Republic, Denmark (west), France, Germany, Greece, Hungary, Italy, Luxembourg, Macedonia, Montenegro, the Netherlands, Poland, Portugal, Romania, Serbia, Slovakia, Slovenia, Spain and Switzerland.

ENTSO-E consists of detailed data on the production, consumption and exchange of electricity for every member country.[36] The largest source of electricity production in the area is the nuclear power plant, which generates around 26 per cent of total electricity. Renewable energy sources (RES) such as hydro, wind and biomass constitute 30 per cent of energy sources, while the rest is generated by solar, tide and geothermal sources.

A number of research studies have examined the risks and barriers to extending the grid's infrastructures. One study performed a risk analysis on the super grid system in the European market and came to the conclusion that integration for energy production has two main risks.[37] These are: not securing a constant electricity supply due to the fluctuating nature of RES, as well as change in demand growth and of not being cost-effective, given the development of technologies of alternative options.

The SUSPLAN project provided insights into the barriers and risks when examining the integrated grid investments that are required to integrate more RES between the years 2030 and 2050.[38] Barriers that prevent integration relate to economic, social and technical aspects, and include:

- *Economic*: this refers to the difficulty in applying a cost–benefit method that economically allots grid investment costs between countries.
- *Social*: the social factor refers to challenges in interconnected grid development due to social factors. It may be that a large number of authorities may be involved in the planning and installation process, which makes the process complex and difficult to handle. Moreover, the public's resistance to extensions in the grid may be present due to potentially adverse impacts on their property, health and the environment.
- *Technical*: the technical factors include limited harmonisation between the five synchronous regions in the grid due to dissimilar operating codes and ineffective coordination between TSOs.

On 25 February 2016 the European Commission issued a progress report on 'Making Europe's electricity grid fit for 2020', mentioning 'achieving a 10% electricity interconnection target', and saying it is the minimum required to end the electricity isolation of some member states and boost competition to bring down prices. Today it has become an essential part of the energy transition to a low-carbon economy. EU Climate and Energy Commissioner Miguel Arias Cañete said at the European Parliament in March: 'There cannot be an increase in renewables without an increase in interconnections.' A 10 per cent interconnection goal means that each member state must have interconnection capacity that is equal to at least a tenth of its installed electricity production capacity. There are different methodologies of measuring interconnection capacity yielding similar results; a 15 per cent interconnection target is in line with that of ENTSO-E. ENTSO-E expects that interconnection capacities should on average double by 2030.

A number of specific schemes have been proposed to create super grids of varying extent within Europe. These include:

- Baltic Energy Market Interconnection Plan involving Denmark, Estonia, Finland, Germany, Latvia, Lithuania, Norway, Poland and Sweden;[39]
- Europagrid, proposed by Europagrid Limited to link various European countries including the United Kingdom, Ireland, the Netherlands, Belgium, Germany and Norway;
- North Sea Offshore Grid, an active proposal by the European Commission, first proposed in November 2008 as a building block towards a Europe-wide super grid[40] involving Germany, the United Kingdom, France, Denmark, Sweden, the Netherlands, Belgium, Ireland and Luxembourg;
- *Low Grid*, proposed by Greenpeace to link the countries of Central Europe, particularly Germany, the Netherlands, Belgium and France;[41]
- *High Grid*, proposed by Greenpeace to link Europe and North Africa, emphasising the installation of solar power in the South of Europe;[42]
- ISLES, an active proposal, at feasibility stage as of September 2011, to link Scotland, Northern Ireland and Ireland with off-shore renewable energy generation.[43]

Super grids (Europe–MENA)

There is high potential for generating renewable energy at present in the neighbouring countries included in the Middle East and North Africa (MENA) region. That is why there has been a renewed interest in generating electricity in MENA and exporting it to European regions. As Table 11.2 shows, an industrial project has been launched to study the practicality of meeting energy demand by generating electricity from this region.

Below are the main objectives of the projects that have been undertaken for the development of integrated electricity networks between Europe and MENA countries:

1   attain energy supply security and reduce climate change risk by sourcing electricity from suitable sites using renewable energy sources that are abundant;[44]
2   transmit generated electricity over long distances to load centres using HVDC transmission lines;[45]
3   ensure greater integration of renewable sources of energy with the present infrastructure by the introduction of an HVDC transmission connection between different countries;[46]
4   facilitate a continuous development of different types of renewable energy sources to improve energy efficiency in all the respective countries and limit the harmful impact of power generation on the environment;[47]
5   develop political, economic, legal and regulatory conditions for a unified trans-Mediterranean renewable electricity market, with the aim of providing reliable and affordable electricity to all countries involved;[48]
6   improve energy supply and economic development for countries south of the Mediterranean Sea by developing framework conditions for electricity export and trading while considering the diversity and priorities of the various countries.[49]

Table 11.2 Projects on the interconnected electricity network between Europe and MENA

Objective	Duration	Region	Timescale	Renewable source	Transmission technology	Type of analysis	Project/ Reference
Generate about 17% of electricity consumption in EU from renewable energy sources in the MENA region by 2050 using HVDC lines that cross the Mediterranean Sea.	2003–present	EU MENA	2050	Solar, wind, hydro, geothermal and biomass	HVDC and HVAC	Techno-economic	DESERTEC concept [11]
Introduce an interconnected power system between Europe and North Africa that is capable of transmitting electricity from different renewable power plants to load centres while managing the fluctuating renewable supply.	2007–present	EU and North Africa	2050	Solar, wind, hydro and biomass	HVDC and HVAC	Techno-economic	SuperSmart Grid (SSG) [45]
Produce 20GW from renewable energy sources in the Mediterranean countries with 5GW exported to Europe by 2020.	2008–present	EU MENA	2020	Solar, wind, hydro and biomass	N/A	Techno-economic	MSP (Mediterranean Solar Plan) [46]
Provide support to the MSP project by introducing a Mediterranean transmission network to transmit electricity from renewable plants to load centres.	2010–present	EU MENA	2020	N/A	HVDC	Techno-economic	Medgrid [47]

Identifying the gaps in literature

A review of the existing literature sheds light on the fact that the prime motive for the introduction of Europe–MENA super grid projects is to reduce the carbon footprint involved in electricity generation. Another motivation of the combined initiative is the development of an integrated power market.

Due to the fluctuating seasonal demands of electricity in the European region, most of the models that optimise the production and generation of electricity in the European IEN are based on situations that make effective use of resources by balancing load supply and requirements in different countries. The analysis results in the creation of multiple energy pathways and offers benefits for areas with constantly fluctuating demands.

A number of studies have examined the viability of an integrated electrical network between MENA and the European region. Studies have mainly focused on imports of electricity from the MENA region to meet the demand in European countries. Almost no research work has been conducted on the techniques to maximise the supply of electricity between the two regions so as to effectively meet the demand.

Modelling super grids

Super grids can be modelled in different ways. The most recent techniques that have been identified in research studies are presented below:[50,51]

- Three-Phase AC technology 50Hz (AC grids) with voltages > 400KV (750KV, 1,000KV);
- HVDC with network-controlled converters (LCC-HVDC, HVDC classic);
- Three-Phase AC technology with reduced frequency (AC grids 16 2/3Hz) with voltages > 400KV;
- HVDC with self-commutated converters (VSC-HVDC).

The HVDC systems are cost effective as compared to AC systems, due to the relatively low costs of the transformer systems. HVDC systems have obvious advantages over AC systems when it comes to long distance transmission. This makes them ideally suited for use in super grids that source electricity from different renewable sources and distribute it to multiple countries over a long distance. Direct benefits include improved working conditions, super power flow control, and a robust platform for increasing supply in the future that can include both renewable energy parks and conventional power plants.

Converter stations act as a solid foundation for an effective HVDC distribution system. Two kinds of converter technologies are Self-Commutated Voltage Source Converters (VSCs) and Line Commutated Current Source Converters (CSCs).

VSCs that utilise the Insulated Gate Bipolar Transistor (IGBT) valves and the Pulse Width Modulation (PWM) method can result in a sinusoidal AC voltage that is completely controllable when it comes to the phase and magnitude of an AC wave. VSCs do not have any reactive power demand. This is unlike the CSC system. They can also substitute the reactive power with an AC grid.

VSCs have the ability to quickly control the active power exchange through control of the phase angle of the voltage. They are also able to achieve this through managing the magnitude of the VSC voltage which is independent of the DC power transmission.

A number of attempts have also been made to formulate an idea of the meshed grids through CSC or HVDC technology. That being said, the high amount of complexity that was involved in the project thereby limited the assessment to just three nodes.[52]

By contrast, the VSC-HVDC offers an adequate condition for a multi-terminal system that forms the basis for super grid modelling, since the total number of modes and the type grid topology that is used is not limited, as is the case for VSC-HVDC.

Multi-Terminal Direct Current (MTDC) systems have been used in a practical power system since 1987 after it was installed in a third terminal in Corsica to a link that ran between Italy and Sardinia.[53]

Currently the POWERGRID Corporation of India (PGCIL) is in the process of installing a 6,000MW (+/– 800kV) HVDC multi-terminal system that has an approximate length of 1,728km. The terminal will run from the north-eastern region to Agra and will consist of a rectifier station in three regions in India. The project is the first of its kind in India that utilises VSCs.

Renewable sources of energy have an intermittent nature that can prove to be harmful for the normal functioning of a large and interconnected grid. The reason for this is the fluctuation in the system that endangers the system's stability.

A DC super grid that is based on the VSC system has the benefit that it makes it possible for non-conventional energy from multiple energy sources to feed an electric current into a common DC super grid. That offers all the respective participants in the different countries access to a stable and reliable source of electricity.

11.8 Existing models and gaps

A number of dispatch and investment models are well established. Some examples include LIMES[54,55] – a multi-scale power system model of 20 regions connected by 32 transmission corridors that integrate optimal investment allocation in grid and generation capacities into a single optimisation framework; and MTSIM[56] – a zonal electricity market simulator with one node per country that determines hourly market clearing prices for a whole year.

There are also electricity system models such as LEAP, GTMax, WASP, Wilmar, EMPS, RAMSES, AURORAx, EMCAS, BALMOREL, MARKAL/TIMES, Mesap Planet, AEOLIUS, IKARIUS, UPLAN, SIVAEL, EnergyPLAN, PLEXOS and EuroDyS. However, each dispatch or investment model or electricity model has different temporal and spatial resolutions, and different capabilities: they can be used to determine the optimal size and timing of new investments in generation and transmission, to perform cost–benefit and market benefit analysis, to develop chronological load forecast series and to project short- to long-term capacity adequacy. From these only a few have a long- and short-term time frame, are free and include an hourly time step. Some have technical restrictions in terms of spatial and temporal coverage.

To my knowledge, a single model with high geographic and temporal resolutions that represents every node of the European electricity transmission network for every half hour and with spatial and temporal resolution that not only represents every node, but also is able to look inside the node, doesn't exist. The main difficulty with such models is to populate each node with supply and demand data. Modelling of interconnected electricity networks has been mostly cross-border centred, with countries or groups of countries represented as a single node and no inclusion of the transmission grids of individual countries.

11.9 Proposed methodology and initial results

I am going to describe here a model I designed to study an interconnected regional system with high renewable energy sources penetration. Note, the model has been developed over the years starting in 2010. One of the objectives was to assess the reliability of the system following interconnection plans and the management of energy demand under variable supply (when large quantities of RES are included) and the need for energy storage. By increasing the spatial resolution of the current node to node methodology by modelling individual generating plants and aggregating the national HV transmission grids in order to address the issues of congestion and renewable energy curtailment in national grids, we can also account for congestion and renewable energy curtailment in national grids that impacts electricity exchanges between countries. Each country within each region is represented by a single node that is interconnected by links with transmission capacity values. The total dispatch costs have been minimised as the sum of variable generation costs and cross-border transmission costs. At each node and each time step, the sum of electricity generation of all energy sources and electricity imports must be equal to the sum of electricity demand, storage and electricity exports. The inputs required for such a model are: net generation capacities for each energy source at each node; cross-border transmission limits; weather data at each node; and electricity production costs per energy source. The outputs consist of electricity dispatch costs; cross-border electricity flows; and energy source generation at each node.

As an initial testing exercise the model has been applied in the EU and West Africa regions. For the EU, the publicly available data from ENTSO-E statistics has been used, while for West Africa, the WAPP-ICC Statistics. Some of these results are presented in Spataru *et al.* (2016).[57]

Figures 11.6 and 11.7 show the net transfer capacity for the EU and WA.

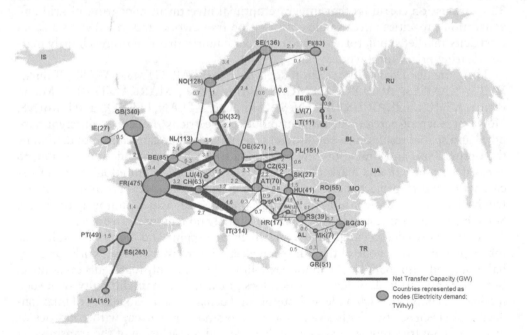

Figure 11.6 Net transfer capacity in the EU.

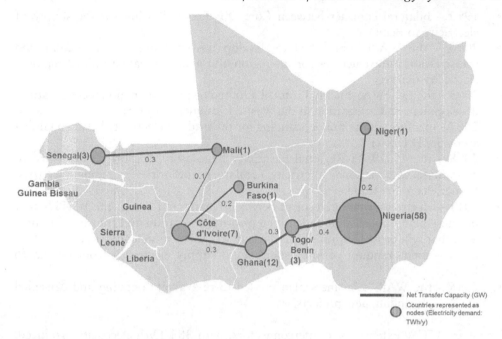

Figure 11.7 Net transfer capacity in West Africa.

In systems with high growth in demand that require significant investments in the short term, such as those of emerging economies in West Africa (Figure 11.9), deployment of high levels of wind and solar power can be carried out in line with grid extensions, so long-term investment plans that consider additional system flexibility are important.

We have also to look at interconnection extensions in 2030 for the WA region (Figure 11.8). Trading of electricity in West Africa currently happens in 9 out of 14 countries in the region. The trading between these countries is through bilateral contracts, trade agreements and memorandum of understanding. In an effort to tackle these challenges, which are crucial for the development of a country's economy, the Economic Community of West Africa States (ECOWAS) established the West African Power Pool (WAPP). The WAPP is made up of 14 out of 15 ECOWAS member countries but, unlike the ENTSO-E in Europe, the 26 members of the WAPP are not exclusively transmission system operators but also include generation and distribution utility companies. The objective of the WAPP is to establish an interconnected electricity grid for the region that will provide reliable electricity through the development of power plants and transmission infrastructures, and also create a unified regional electricity market.

The time frame of electricity trade in West Africa is as follows:

* 1992 – bilateral contract between Nigeria and Niger for the supply of electricity to Niger;
* 1997 – bilateral contract between Côte D'Ivoire and Burkina Faso for the supply of electricity to Burkina Faso; bilateral contract between Nigeria and Benin for the supply of electricity to Benin;

- 1999 – bilateral contract between Côte D'Ivoire and Benin for the supply of electricity to Benin;
- 2000 – the WAPP established to develop power plants and interconnected transmission infrastructures for the region and also to create a unified regional electricity market;
- 2003 – the ECOWAS Energy Protocol adopted to promote regional cooperation, development and integration in the West Africa energy sector;
- 2007 – memorandum of understanding for the trade of electricity between Ghana and Côte D'Ivoire;
- 2008 – the ECOWAS Regional Electricity Regulatory Authority established to coordinate with national electricity regulatory authorities in regulating cross-border electricity trade;
- 2011 – electricity trade agreement between Ghana and Benin; the WAPP's first business plan published and identified priority generation and transmission projects that will provide affordable electricity to two or more countries;
- 2013 – memorandum of understanding to supply electricity from Ghana to Benin;
- 2015 – the WAPP's business plan revised to re-evaluate ongoing and cancelled projects, including new projects.

There are 15GW extensions to interconnectors, with 383 TWh electricity produced, with 15 per cent of it (57TWh) exchanged between countries. Nigeria, Ghana, Côte d'Ivoire and Guinea are the major exporters due to large cheaper base load generation (gas and hydro).

Diesel plants are dispatched to ensure peak demand is met, thus resulting in a significantly higher marginal cost of production.

Battery or pumped hydro storage charged by the utility PV during day time could replace diesel plants during peak time to reduce cost.

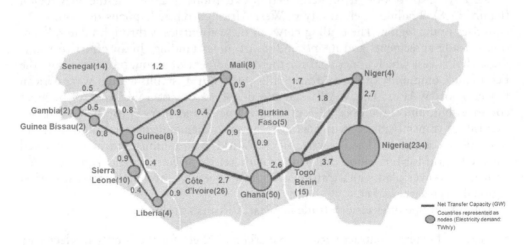

Figure 11.8 Transmission capacity and electricity demand.

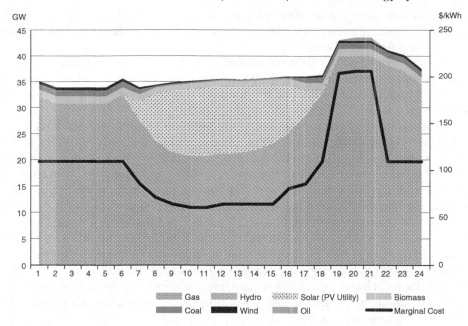

GW
$/kWh

Gas Hydro Solar (PV Utility) Biomass
Coal Wind Oil Marginal Cost

Figure 11.9 Electricity dispatch on a dry day (peak day March).

The methodology is currently expanded to include market coupling. One of the challenges currently facing market coupling in Europe is congestion management due to increasing electricity supply, especially from RES; this challenge holds an opportunity for the West Africa region to account for sufficient capacities as extensions to national and cross-border transmission infrastructures are being planned.

For the moment, economic market constraints are not included. The price differences can be significant. Ouriachi and Spataru[58] have shown that, between July 2012 and June 2013, NordPool benefited from the lowest electricity prices (thanks to their extensive hydraulic capacities) with an annual average of 34.7€/MWh, while Italy paid the highest price (due to its lack of interconnection and use of fossil fuel-based electricity mix) with an annual average of 67.2€/MWh (almost twice the Nordic price).

The methodology can be applied to other regional case studies, such as South America or the MENA countries.

11.10 The global energy grid

There has been great integration and extension of the electric supply system ever since the installation of the first public lighting by the Godalming Borough Lighting Committee in 1881 and the first international link in 1914.[59] During the 1930s, Buckminster Fuller had proposed a global grid with the idea that when the world is connected through an electrical energy grid, it will allow the integration of renewable resources located in different locations to meet the energy needs of the world.[60]

In the year 1960, Bucky saw that the energy needs could be transferred cost effectively at a distance of about 1,500 miles. He had designed a grid that would help in carrying the loads while taking into account population centres and renewable energy resources as well as the requirements to make the connection. He had also viewed the points where waste might occur and figured out how to divert the waste and feed it to the energy supply to meet shortfalls.

Bucky also examined the Dymaxion Map in both the folded and unfolded form. In the traditional folded form, he showed the world to be one huge island contained in one world ocean. A big stumbling block was of course the politics that surrounded national boundaries.

An example can be given in this context of the Straits of Gibraltar that have a very strong current. The current is strong enough to supply affordable electricity to millions of people even if only a small portion of the total capacity was tapped through large, low RPM turbine equipment that will not hinder any movement of marine life.

However, in the case where Morocco and Spain agree to harness the lost energy at the junction point, there is a problem in terms of where to prioritise the supply of energy.

This is a question that needs to be looked into during the planning stage. Payments that a country requires for passing electricity through its borders should also be looked at. Such charges can greatly increase the cost of the energy supplied to the end user.

That being said, the development of a super grid is happening at a slow pace. Most of the development has occurred in limited areas and within the borders of first world countries including in North America and Europe. And even there the progress has occurred in a haphazard manner.

However, the policy planners and engineers in the first world countries have done an excellent job of implementing and planning large-scale power systems for meeting demand decades ahead into the future.

But there is more that needs to be done for optimal utilisation of clean sources of energy. Some of the questions that should be catered for include: How to optimally utilise wind turbines to ensure that the power needs are met efficiently? How to implement renewable sources of energy with success and economical efficiency? And lastly, how to efficiently simulate scenarios to depict real world supply of and demand for energy?

11.11 How to unlock the potential of renewable energy around the world

Development of international and regional super grids and the greater usage of HVDC technologies is one way to unlock the true potential of a renewable energy future. Although renewable energy such as solar and wind are intermittent at a specific location, when looked at on a global scale it can offer a secure and stable source of electricity.

At the moment a regional super grid is not on the cards in many countries, even in countries such as the US. Implementation of a global super grid is even harder to imagine. Implementation of the super grid will require making significant investment in terms of infrastructure and policy requirements. This in itself comes with its own sets of challenges. Some of these include maintenance and management of a large energy flow and geopolitical consideration.

Investments in a highly effectual and efficient globally connected super grid will go a long way towards limiting the carbon footprint. This can also offer fruitful results at a national level as well. In some parts of the world, especially the US, a lot has been done in this regard, including the addition of the TresAmigas converter station and multiple projects that use HVDC to integrate the Western, Rasyern and ERCOT grids. Atlantic Wind Connection will bring offshore power to the cities via New England.

Most of the countries in Europe, led in part by Germany, have contributed considerable funds to upgrading the transmission system that would make a super grid a reality. However, a lot more work needs to be done at present to achieve the goal of unlocking the potential of renewable energy sources and eliminating carbon footprints from the environment.

China has also invested similarly in HVDC transmission to take advantage of its large hydroelectric resource, and open up the latest opportunity throughout the country. An article published on Reuters in 2012 gave a snapshot of some progress that has been made with regard to the development of a super grid in different regions and countries around the world.[61] The Global Energy Network Institute is also promoting the concept but on a global scale. And this is important. The prospect of achieving the goal of 100 per cent renewable energy requires cooperation at a global level.

Notes

1 IEA Heat Pump Centre. Accessed May 2016 from www.heatpumpcentre.org
2 N. Hewitt, M. Huang, M. Anderson and M. Quinn (2011). Advanced air source heat pumps for UK and European domestic buildings. *Applied Thermal Engineering*, 31 (17–18), 3713–3719.
3 Mitsubishi Electric Ecodan Newsletter (2008). *Ecodan Newsletter*, Issue 2.
4 N. Kelly and J. Cockroft (2011). Analysis of retrofit air source heat pump performance: Results from detailed simulations and comparison to field trial data. *Energy and Buildings*, 43, 239–245.
5 IEA Heat Pump Centre. Accessed May 2016 from www.heatpumpcentre.org
6 Z. Yang, G. Pedersen, L. Larsen and H. Thybo (2007). Modelling and control of indoor climate using heat pump based floor heating system. *33rd Annual Conference of IEEE Industrial Electronics Society (IECON), Taipei, Taiwan, 5–8 November 2007*.
7 A. Schijndel and M. de Wit (2003). Advanced simulation of building systems and control with Simulink. *8th International IBPSA Conference, Eindhoven, Netherlands, 11–14 August 2003*.
8 N. Kelly and J. Cockroft (2011). Analysis of retrofit air source heat pump performance: Results from detailed simulations and comparison to field trial data. *Energy and Buildings*, 43, 239–245.
9 A. Schijndel and M. de Wit (2003). Advanced simulation of building systems and control with Simulink. *8th International IBPSA Conference, Eindhoven, Netherlands, 11–14 August 2003*.
10 Mathworks (2015). Accessed February 2016 from http://uk.mathworks.com
11 Z. Yang, G. Pedersen, L. Larsen and H. Thybo (2007). Modelling and control of indoor climate using heat pump based floor heating system. *33rd Annual Conference of IEEE Industrial Electronics Society (IECON), Taipei, Taiwan, 5–8 November 2007*.
12 A. Schijndel and M. de Wit (2003). Advanced simulation of building systems and control with Simulink. *8th International IBPSA Conference, Eindhoven, Netherlands, 11–14 August 2003*.
13 G. Rogers and Y. Mayhew (1992). *Engineering Thermodynamics: Work & Heat Transfer*, 4th edn. Harlow: Pearson Education.
14 A. Rabl (1988). Parameter estimation in buildings: Methods for dynamic analysis of measured energy use. *Journal of Solar Energy Engineering*, 110, 52–66.
15 J. Dormand and P. Prince (1980). A family of embedded Runge-Kutta formulae. *Journal of Computational and Applied Mathematics*, 6 (1), 19–26.
16 Mathworks (2015). Accessed February 2016 from http://uk.mathworks.com
17 M. Barrett and C. Spataru (2012). *A Dynamic Energy Agents-Based Model (DEAM) – Falcon Project*, November 2012. Report to the Energy Savings Trust. Accessed January 2016 from www.ucl.ac.uk (energy-models/models/deam)
18 C. Spataru and M. Barrett (2015). *DEAM – A Scalable Dynamic Energy Agent Model for Demand and Supply*. 17th UKSIM-AMSS International Conference on Modelling and Simulation. Accessed May 2015 from http://uksim.info/uksim2015/data/8713a195.pdf

19 Centre of Sustainable Energy. *Display Energy Certificate Data.* Accessed May 2016 from www.cse.org.uk/projects/view/1259#Display_Energy_Certificate_data
20 UK Marketing Database. Accessed May 2016 from www.cd-rom-directories.co.uk/contents/en-us/d317_United_Kingdom_Marketing_Data.html?gclid=CN3guMOQsq4C FaQmtAodqUm_Og
21 C. Spataru and M. Barrett (2015). *DEAM – A Scalable Dynamic Energy Agent Model for Demand and Supply.* 17th UKSIM-AMSS International Conference on Modelling and Simulation. Accessed May 2015 from http://uksim.info/uksim2015/data/8713a195.pdf
22 Ibid.
23 Ibid.
24 M. B. Blarke and B. M. Jenkins (2013). SuperGrid or SmartGrid: Competing strategies for large-scale integration of intermittent renewables? *Energy Policy,* 58, 381–390. Accessed January 2015 from http://linkinghub.elsevier.com/retrieve/pii/S0301421513002176
25 C. Macilwain (2010). Supergrid. *Nature,* 468, 624–625.
26 European Commission (2011). *Energy Infrastructure – Priorities for 2020 and Beyond.* A Blueprint for an Integrated European Network.
27 A. Battaglini, J. Lilliestam, A. Haas and A. Patt (2009). Development of supersmart grids for a more efficient utilisation of electricity from renewable sources. *Journal of Cleaner Production,* 17 (10), 911–918.
28 D. Van Hertem and M. Ghandhari (2010). Multi-terminal VSC HVDC for the European supergrid: Obstacles. *Renewable and Sustainable Energy Reviews,* 14 (9), 3156–3163.
29 D. P. Kaundinya, P. Balachandra and N. H. Ravindranath (2009). Grid-connected versus stand-alone energy systems for decentralized power: A review of literature. *Renewable and Sustainable Energy Reviews,* 13 (8), 2041–2050.
30 Adapted from the report: *Gobitec and Asian super grid for renewable energies in Northeast Asia* by S. Mano, O. Bavuudorj, S. Zafar, M. Pudlik, V. ÜLCH, D. Sokolov and J. Young Yoon (2014). Accessed from www.energycharter.org/fileadmin/DocumentsMedia/Thematic/Gobitec_and_the_Asian_Supergrid_2014_en.pdf
31 K. Komoto, N. Enebish and J. Song (2013). Very large scale PV systems for north-east Asia: Preliminary project proposals for VLS-PV in the Mongolian Gobi Desert. *IEEE 39th Photovoltaic Specialists Conference (PVSC), 2013.*
32 IEA Heat Pump Centre. Accessed May 2016 from www.heatpumpcentre.org
33 *UCPTE-UCTE: The 50 Year Success Story – Evolution of a European Interconnected Grid.* Brussels, Belgium: ENTSO-E – European Network of Transmission System Operators for Electricity. Accessed May 2016 from www.entsoe.eu/fileadmin/user_upload/_library/ publications/ce/110422_UCPTE-UCTE_The50yearSuccessStory.pdf
34 *Union for the Coordination of the Transmission of Electricity (UCTE).* Belgium: ENTSO-E – European Network of Transmission System Operators for Electricity. Accessed May 2016 from www.entsoe.eu/news-events/former-associations/ucte/Pages/default.aspx
35 *ENTSO-E at a Glance.* Belgium: ENTSO-E – European Network of Transmission System Operators for Electricity. Accessed May 2016 from www.entsoe.eu/Documents/ Publications/ENTSO-E%20general%20publications/entsoe_at_a_glance_2015_web.pdf
36 *Statistical Database.* Belgium: ENTSO-E – European Network of Transmission System Operators for Electricity. Accessed May 2016 from www.entsoe.eu/data/data-portal/Pages/ default.aspx
37 M. E. Torbaghan, D. V. L. Hunt and M. Burrow (2014). Supergrid: Projecting interconnection capacities for the UK. *Proceedings of the Institution of Civil Engineers – Engineering Sustainability,* 167 (6), 249–263.
38 *The DESERTEC Concept.* Germany: DESERTEC Foundation. Accessed May 2016 from www.desertec.org/concept
39 House of Commons Energy and Climate Change Committee (2011). *A European Supergrid, Seventh Report of Session 2010–12, Vol. 1.* Accessed from www.publications.parliament. uk/pa/cm201012/cmselect/cmenergy/1040/1040.pdf
40 *Communication from the Commission to the European Parliament, the Council, the European Economic and Social Committee and the Committee of the Regions – Second Strategic Energy Review: An EU Energy Security and Solidarity Action Plan* {SEC(2008) 2870} {SEC(2008) 2871} {SEC(2008) 2872} (PDF). European Commission. November 2008: 4–6. Accessed January 2010.

41 *Battle of the Grids*, Greenpeace International, published 18 January 2011, accessed October 2011.
42 Ibid.
43 House of Commons Energy and Climate Change Committee (2011). *A European Supergrid, Seventh Report of Session 2010–12, Vol. 1.* Accessed from www.publications.parliament. uk/pa/cm201012/cmselect/cmenergy/1040/1040.pdf
44 Mathworks (2015). Accessed January 2016 from http://uk.mathworks.com
45 Ibid.
46 Z. Yang, G. Pedersen, L. Larsen and H. Thybo (2007). Modelling and control of indoor climate using heat pump based floor heating system. *33rd Annual Conference of IEEE Industrial Electronics Society (IECON), Taipei, Taiwan, 5–8 November 2007.*
47 Ibid.
48 Ibid.
49 Ibid.
50 J. J. Grainger and W. D. Stevenson (1994). *Power System Analysis*, Vol. 621. New York: McGraw-Hill.
51 R. C. G. Teive, E. L. Silva and L. G. S. Fonseca (1998). A cooperative expert system for transmission expansion planning of electrical power systems. *IEEE Transactions on Power Systems*, 13 (2), 636–642.
52 P. Meisen and F. Beg (2014). *Super Grid Modelling and Risk Assessment for the Indian Subcontinent.* Accessed February 2015 from www.geni.org/globalenergy/ library/technical-articles/transmission/smart-grid/supergrid-modelling-risk-assessment-indian-subcontinent-farhan-beg.pdf
53 A. Kumar (2000). An efficient super grid protocol for high availability and load balancing. *IEEE Transactions on Computers*, 49 (10), 1126–1133.
54 M. Haller, S. Ludig and N. Bauer (2010). *Fluctuating renewable energy sources and long-term decarbonisation of the power sector: Insights from a conceptual model.* The International Energy Workshop. Stockholm, Sweden.
55 M. Haller, S. Ludig and N. Bauer (2012). Decarbonization scenarios for the EU and MENA power systems: considering spatial distribution and short term dynamics of renewable generation. *Energy Policy* 47, 282–290.
56 A. Zani, G. Migliavacca and A. Grassi (2011). *A scenario analysis for an optimal RES integration into the European transmission grid up to 2050.* 8th International Conference on the European Energy Market (EEM) (pp. 401–406).
57 C. Spataru, T. Adeoye and R. Bleischwitz (2016). Future perspectives on policy instruments and market coupling for integration of RES-e in regional supergrids. *BIEE Conference Innovation and Disruption: The Energy Sector in Transition*, Oxford, 21–22 September.
58 A. Ouriachi and C. Spataru (2015). *Integrating regional electricity markets towards a single European market.* IEEE Xplore.
59 *Godalming and Electricity.* Godalming Museum. Accessed May 2016 from www.godalming museum.org.uk/index.php?page=1881-godalming-and-electricity
60 R. D. Hoffman (2000). *The Potential Impact and Importance of R. Buckminster Fuller's Vision of a Global Energy Grid.* Accessed May 2016 from www.animatedsoftware.com/ geni/rh2000ge.htm
61 J. Kemp (2012). *Super-grids, Mass Blackouts and Clean Energy.* Reuters. Accessed May 2016 from www.reuters.com/article/2012/11/16/us-column-kemp-supergrids-id USBRE8AF0XA20121116.

12 Problems/gaps/research questions and issues

12.1 Data-related problems

A time and cost constraint is linked with organising and analysing raw data relating to wind and solar farms. To gather the required data, we need to obtain permission from the respective energy companies. This will require the signing of different documents including a non-disclosure and confidentiality agreement to gather the required data.

The process could take months or even years. If we have access to the knowledge and data-sharing platform required for renewable energy, it could make our research much easier and great progress could be made.

12.2 Modelling-related problems

Another problem relates to modelling of the data to capture the dynamics at a greater level of detail and integrating the models. The degree of integration requires selecting an integrated model versus a loose coupling model. This all depends on the particular requirements of the systems.

Remember that no one model can capture each and every dynamic, giving a complete picture of the reality. What's best is to select a model that gives a 'good enough' picture of the overall load requirement and that supplies answers in the best manner in view of the complexity of the resources and systems in space and time.

In my opinion, computational general equilibrium (CGE) models should not be used for making projections in the long term despite this being the current practice. Some would say that this is necessary as the parameterisation and production functions are not ideal for projections. However, I believe this is not true and that a CGE model should be used for a short time period as it includes dynamics easily in such a situation.

Most of the existing models are for making policies and exploring scenarios. However, indirectly they are suitable for implementations and planning. Certain approaches used by industry specialists and consultants are more applicable compared to academics.

For instance, dynamic stochastic general equilibrium (DSGE) models that are dynamic in nature are more suitable for consultants compared to academicians. This is despite the fact that the DSGE model is similar to the CGE model.

But when looked at in detail, CGE is more apt for focusing on medium-to-long run macroeconomic situations, while DSGE can capture fluctuations in different business cycles that allow random variations in capturing uncertainty that include the integration of business cycles.

Part III

Energy policy, markets and geopolitics

Prologue

The energy system is currently undergoing considerable changes and slowly transitioning towards the development of new sources that are efficient and reliable. Part III deals with energy policy, markets and geopolitics, with the focus being on global energy policies and their implementation, as well as effect.

Part II

Energy policy, markets and geopolitics

13 Energy policies

What our past teaches us

Definitions (*Oxford Dictionaries*[1]; *UK Parliament*[2]):

> *legislation*: laws that must be complied with in order to remain within the legal boundaries of the country;
>
> *policy*: a course or principle of action adopted or proposed by an institution or organisation;
> *regulation*: a specific rule or directive that is enforced by regulators and supports the requirements of legislation.

An 'energy policy' denotes a system that has been put in place in which any agency (typically state-owned organisations) acts to encourage the promotion, distribution and, eventually, consumption of energy. This is why an all-encompassing energy policy could potentially comprise different laws, agreements (both national as well as international) and incentives for energy development business concerns, along with guidelines for energy conservation as well as other objectives of public polity related to energy.

The instruments of governance may be broadly classified as flowing from the top down: law and legislation, government spending, as well as taxation policies.

a) *the various tools of governance*: administration and bureaucracy; state institutions;
b) *informal*: information tools and instruments, networks and authorities.

Government policies can be collectively described as:

a) *monetary policies*: subsidies, loans, grants, aid, rebates, etc.;
b) *market-oriented policies*: market transformations, trading systems, etc.;
c) *regulatory policies*: standardisation, rules and regulations;
d) *information-based policies*: R&D, training and education, advice, advocacy, protracted campaigns, etc.

Overall climate policies include the following:

a) *international*: international treaties, agreements and protocols, e.g. Kyoto Protocol, COP 21[3], EU directives to all member states;
b) *national*: rules and regulations, legislation, national policies such as carbon budgets, Climate Change Act in the United Kingdom;

c) *local*: government actions at the local level and various multi-tiered programmes such as the Islington Sustainable Energy Partnership (ISEP) and the Camden Climate Change Alliance (CCCA).

13.1 Summary of energy policies: a global perspective

The increased focus on creating sustainable energy policies all over the world dates back to the last years of the previous century when the world was gripped by an unparalleled rise in petroleum, oil and lubricants (POL) prices. This occurred when OPEC (Organization of Petroleum Exporting Countries) member states opted to reduce their oil output drastically. Refineries anticipated a return to normalcy once OPEC increased their production again, and as a result used up their reserves. However, when OPEC refused to jack up its output again, there was a severe POL products shortage, leading to a hitherto unseen rise in oil prices.[4]

In the United States, the 1999–2000 winter was particularly harsh and made even seaborne transport of home heating oil difficult for New England (an area in the north-eastern United States that comprises the states of Connecticut, Maine, Massachusetts, New Hampshire, Rhode Island and Vermont). The resultant surge in prices also affected diesel as well.

The demand for mixing components used in the manufacture of reformulated gasoline (a form of vehicle fuel that helps achieve cleaner air goals) far outstripped its supply and as a direct result led to the increase in the price of gasoline to $2 per gallon. Meanwhile disturbances in oil-producing countries in South America and the Middle East led to shortages that in turn led to a slow but steady increase in prices.

The prolonged period of short supply of fuel products along with their higher prices was reminiscent of the oil crisis of the early 1970s and gave an entire generation of Americans the idea of how important an uninterrupted supply of fuel (along with its price stability) is towards maintaining not only their economy but their very lifestyles as well. As a direct result, the US government quickly developed a comprehensive energy policy so as to ensure that all such unforeseen eventualities are avoided in future through better planning.

This is why today, whenever EU governments experience a surge in POL prices that they pass on to the end consumers, the respective governments of those countries are severely criticised by the stakeholders concerned for not having a comprehensive energy policy to deal with the problem.

After studying the discussions by Mamberger,[5] we can deduce that the oil embargo enforced by various Arab nations had a major impact on the formulation and implementation of energy policies the world over. Let us first discuss exactly what kind of impact the embargo (that the Arab countries had imposed on certain other nations of the world) had on energy policies globally and how it helped change them.

13.2 Traditional energy policies before the 1973 oil crisis

Before the 1973 crisis the private sector, acting on the principles of free market enterprise in the United States, supplied the whole nation with an almost unlimited supply of inexpensive energy. The US itself has plenty of domestic fossil fuels of its own that include oil, natural gas and coal. Additionally, the country also was able to acquire cheap imported gas and oil whose development and supply was entirely in the hands of the private sector. The few government policies that were implemented were done so as to ensure the stability of the pricing mechanism as well as to safeguard the interests of the energy production companies.

Therefore it can be said that energy prices prior to the crisis of 1973 were kept deliberately low through a combination of actions by both the federal government and the governments of the various states that comprise the US. The core purpose was to increase consumption without any emphasis whatsoever on efficiency and protection (of the end consumer), as well as energy liberation of the country at the national level.

This is why the country was entirely unprepared for the 1973 crisis that brewed in the Middle East so unexpectedly and exposed the country's energy supply vulnerability, and what was hitherto a domain controlled entirely by the government as well as the private sector became an issue of great interest to the general public, and as such policymaking too slipped into the public domain.

The pre-1973 energy policy of the US was centred on certain points:

The main principles of the policy

The private sector had full permission to participate in both the exploration as well as production of energy courtesy of the Mineral Leasing Act of 1920. In line with this piece of legislation the only thing that was required from the energy-producing concern was the payment of certain fees at the federal as well as state level. It has been speculated that this type of 'freewheeling' policy gave rise to the famed 'oil barons' of yesteryear.

Taxation policy

The pre-1973 era offered multiple tax breaks to various energy producers, with the core focus being on encouraging both exploration and subsequent production. Indeed, so effective was the tax rule that it led to an over-production glut. This glut forced the various states to commence proration. Proration meant that certain production limits were imposed on the oil magnates and their business concerns so that supply could be maintained at a certain level and, as such, prices would remain stable and the producers themselves would be saved from over-competition and price wars.

Antitrust policies

The government's decision to use portioning was in direct conflict with antitrust rules whose purpose was to actually encourage competition in line with the capitalist ideals on which the nation's economy was based. However, due to the heavy demand for energy during the Second World War, the government had no choice but to abandon its antitrust efforts.

Free market economic principles

The energy policies that were in place pre 1973 were heavily tilted in favour of the private sector, with the government aiming for a minimalist approach to the energy sector. This was because the government wanted the sector to operate purely as a 'free market', and so the government opted to keep out of it to a great extent so as to ensure that the private sector was able to fulfil all the energy demands of the country, on its own.

Coal as fuel

The plentiful supply and low price of this fuel meant that the country depended purely on coal for almost two-thirds of its energy usage until at least the late 1920s. This 'coal glut' meant that there was always a surplus, and therefore the business concerns providing coal had to do whatever they could to ensure their product was sold, even if it meant near-constant price wars. Consequently, only a few selected coal supply companies were in a position to reap good dividends and profits. To curb this over-supply, a series of Bituminous Coal Acts were enacted in the years 1935 and 1937. However, they failed to have the desired impact.

But nevertheless the end of the Second World War produced a paradigm shift and oil was considered a far better alternative to coal as a source of energy. And this ultimately led to a tapering off of the demand and subsequent usage of coal as an energy source.

Oil

The vast majority of oil production facilities have traditionally been in the Texas heartland, with the term 'Texas oil baron' being a proverbial watchword in popular American folklore. Given that the majority of oil was initially produced from Texas, the national oil policy throughout the 1930s all the way up till the early 1950s was spearheaded by the Texas Road Commission. After the state of Texas had imposed portioning, other states promptly followed suit.

Oil depletion allowance

This allowance was enacted by the US federal government in 1926 and was a key plank in the safeguarding aspect of the nation's energy policies. After the end of the First World War, the United States Geological Survey predicted that the country's oil reserves were going to drop very quickly in the next decade or so. The oil sector responded by treating oil as a depreciable commodity akin to its machinery. In order to keep the industry happy, Congress passed a rule that enabled the various companies operating in the oil sector to remove 2.7 per cent of their gross profit from their taxable income.

Even though the predictions turned out to be completely wrong and vast oil reserves were discovered continuously throughout the 1930s, this oil depletion allowance was not removed on the assumption that it would act as a viable incentive for greater exploration.

Natural gas

Natural gas is a less demanding alternative to oil since it does not require a lot of refining. The three main activities for natural gas are: at-source production, supply, and eventual distribution to the end consumer. To regulate the supply the Natural Gas Act of 1938 was promulgated. This act mandated the Federal Power

Commission as the body vested with the authority responsible for both regulating the construction of the pipeline and the price of the natural gas in the interstate market. As a result, a cheap gas supply was assured; however, its low price was a major hindrance to its development.

Nuclear energy

Throughout the 1950s and 1960s, nuclear power was widely perceived as a major source of energy in Western countries. The American Congress formed the Atomic Energy Commission. This body aimed to promote atomic power. However, government ceded the construction of atomic power plants themselves to the private sector.

13.3 Western dominance over oil-rich nations' production facilities

Thanks to US dominance in the field of exploration and production of energy, their companies were the mainstays of fossil fuel exploration even in the oil-producing heartland of the Middle East and Africa. In fact, pre 1973 the oil-producing nations did not have full control over their own resources due to American and European business concerns having taken charge of both production and distribution of their fossil fuel resources.

By 1930 Chevron, Texaco, British Petroleum, Mobil, Exxon, Royal Dutch Shell and other top oil-producing international companies had successfully managed to sign collaborative treaties regarding oil exploration in different countries. This resulted in a steep decline in competition. Aramco, for instance, was and still is in control of most of the Saudi oil reserves to this day. Indeed, so effective was this cooperation that Western multinational corporations (MNCs) were in control of almost the entire world's crude oil production.

Being cognisant of this fact, sovereign oil-producing nations formed a global entity called the Organization of Petroleum Exporting Countries (OPEC) in 1960 to wrest back control of their own natural resources. Originally, OPEC had seven Arab members and six non-Arab nation members. At the time of its formation the US did not see the formation of OPEC as a direct energy threat to itself since it had almost achieved self-reliance in oil production and was importing very minimal quantities indeed – so much so that production was actually higher than demand.

However, the socio-political events of 1973 were a watershed moment for the West as the seven Arab OPEC members decided to use POL as a weapon against Israel as well as the countries that supported it. The nations that supported Israel during the Arab–Israeli conflict had their oil imports reduced by 5 per cent. However, the embargo did not produce the desired result since its main target, the US, simply shifted its imports to Venezuela and Iran, and as a result the embargo was lifted within a year of its imposition. However, it did succeed in eroding support for Israel in some Western European countries as well as Japan.

However, the Arab embargo's main impact was that it showed the collective power of the oil-producing nations. Had the non Arab countries that comprise a part of OPEC followed suit, their demands would almost certainly have been met.[6] Accordingly, OPEC decided to act in tandem, and if one member nation raised prices the others followed suit.

World dependence on fossil fuels

Oil has for the past few decades been the mainstay of the world's energy needs. In fact oil and gas supply over 55 per cent of global energy. Currently the industrialised world relies a lot on imported oil as the US is reportedly importing about 40 per cent of its oil, while both Europe and Japan are reported to be importing over 55 per cent of their oil as well. Western Europe sources its oil from Russia, North Africa and the Middle East, and Japan imports 59 per cent of its oil from Islamic countries. The US, after being at the receiving end of the Arab world embargo in 1973, has made a point of only importing about 10 to 12 per cent of its oil from Islamic countries.

13.4 Energy policy reviews

Energy policies may be reviewed through a spectrum of four broadly different strands of research: one aspect emphasises reviewing both the development as well as the achievements of energy policies as they are implemented in a country, region, state, union or sector;[7,8,9,10,11] the second stream is concerned with the rules and regulations and their consequent outcomes as they relate to specific energy forms (e.g. electricity, gas, oil) globally;[12,13] the third aspect is related to the effects of different types of policy frameworks for energy regulation;[14] and finally the last strand of research is concerned with the comparison of the features of energy governance as they exist in different countries.[15]

Taken in combination, all of these studies reveal certain 'facets' of different energy policies. Here I provide a brief visualisation of world literature on this topic by attempting to develop a spatial and temporal perspective – roughly analogous to a screenshot of the historical and dynamic trends of global energy policy studies.

To make sure it's a comprehensive and accurate review, I have provided examples and analysis of the RES policies in the European Union (both wind and solar) and have also included the electricity coupling markets in Europe. A sound understanding of these policies may help in applying the best practice to other countries and regions all over the world.

A comprehensive energy policy typically has multidisciplinary components that may include but are not restricted to including materials sciences, politics, geological resources, economics, engineering, ecology, biology and law, as well as environmental assessment. Taken together they will help explain why most decision making in this area tends to be both complicated and controversial.

Energy policies are typically a topic of serious concern among individual countries, and as far as the European Union is concerned have quite often generated highly charged debates at the intergovernmental level as these negotiations have generally being aimed at establishing a (standardised) European energy policy.[16] A powerful bond gradually developed between the domestic energy policies of individual countries and the global political environment when The Control of Oil[17] was manipulated by 12 leading members of OPEC to fix world oil prices for their own benefit. This manipulation was what effectively triggered the world global energy crisis of the 1970s, in spite of the fact that US President Carter's administration policies proved to be successful for the overall energy security needs of the US, thanks in part to the fact that he got rid of the influential Joint Committee on Atomic Energy and also went on to issue the Crude Oil Windfall Profits Tax. However, relations between the US government and markets always tend to be uneasy.

In fact US researchers pioneered the main theoretical analytical methods that consist of combinatorial optimisation and decision analysis, as well as econometrics, operational research and simulation analysis, all of which were designed to help resolve energy-related matters in the overall policy decision-making processes post the 1970s era energy crisis period. Among all the models that are currently popular these days, mathematical modelling is the most conservative way to deal with the crisis, followed by the 'historic insight' method.[18] Indeed, energy economics appeared as far back as the early 1910s and lasted for over 20 years as one of energy research's key frontiers.

Energy conservation measures were generally used from the 1960s to the 1980s as a key plank of energy policy so as to improve overall energy security both for individual countries as well as collectively. In addition to the ubiquitous fossil fuels, nuclear power as an energy source was the policymakers' hands-on favourite and formed a 'hard' path to achieve both national energy security and independence.[19] Nevertheless the operation of nuclear power plants is inherently controversial in nature since several accidents have already occurred,[20] with the worst being the Chernobyl disaster in 1986 in the then USSR.

In Sweden the 'mushrooming' (no pun intended) construction of nuclear power plants from the 1960s to the 1970s led to an increasingly acrimonious debate on the overall safety of nuclear power vis-à-vis the environment. These debates in effect forced the government to change its policy from atomic-focused to a more 'diversified renewable scenario' from 1977 to 1994.[21] As of today, Sweden ranks among the leading International Energy Agency (IEA) members globally with regard to low-carbon intensity and high share of renewables in its cumulative energy supply matrix, with strong growth coming from solid biofuels as well as onshore wind projects.[22]

Energy technology research today has reached its peak with the introduction of the Stabilization Wedges Theory[23] as well as other ancillary studies.[24,25] These research studies have been highly successful in providing a diverse array of options – such as overall energy efficiency management, nuclear power, fuel shifting, low carbon technology, carbon capture as well as storage, and power from renewable sources – to keep emissions not only at a stable and acceptable level but also to help address climate change and its resultant issues.

In the last decade of the previous century, energy policymaking became a complex and contentious issue as climate change started taking a toll on the global environment. Energy growth is currently principally driven by population and economic growth rates, and the high level of consumption is chiefly driven by population and economic growth in emerging countries, such as the BRICS nations (Brazil, Russia, India, China and South Africa), whose high consumption of fossil fuels have centred global attention on the energy crisis.

From the 1960s to the 1980s a substantial number of studies were conducted regarding both the substitution and the complementary effects between energy and non-energy elements to understand the impact of tax and/or price on macro-economy fluctuations.[26,27]

In terms of methodology, many energy policy choices have been based on using 'elastic analysis methods'. These include: the Morishima Elasticity of Substitution, Allen-Uzawa Partial Elasticity of Substitution, and the Cross Price Elasticity Theory of Neo-Classicism. Based on the 'business cycle concept', Christensen's Translog Production Function[28] has become the mainstay of technology improvement measurement. In fact Craig's work has also helped generate debate that focuses on the relationship between consumer behaviour and energy conservation when both are seen through the lens of effective communication.[29] In fact, *using* information to convince the public to move towards greater energy conservation by specifically targeting consumer behaviour has become one of the favourite topics in the energy policy discipline.

As an essential element of exhaustible theory, Hotelling's law[30] drew public notice to both the concept as well as the consequence of 'indefinite' fossil fuel consumption. On the other hand, the eventual upper limit price of energy and also the shortage of electricity generation capacity that may be created by highly unreasonable policy approaches may (in the long run) cause myriad difficulties in the national energy security needs of individual nation states.

This is why policies affecting different energy markets have been subjected to rigorous scrutiny. It was mainly to underline the importance of systematic economic analysis. In this perspective the electricity market theory has emerged – a theory that is based mainly on marginal cost analysis. As an expert in this field, Joskow has made a significant addition to the electricity market theory with his research on the 'electricity market capacity mechanisms', demand side management (DSM) effectiveness, power system reliability generation, the capacity regulation mechanism, and industry performance related to plant scale.

Sundquist,[31] on the other hand, proposed the establishment of a single 'unified' federalism guided by the office of the president, that blithely ignores the 'potential losses in efficiency or responsiveness [that are] endemic to the performance of large-scale bureaucracies' the world over; whereas Light[32] discussed the evolution of the state's enhanced role during and after the 1973–1974 energy crisis.

Comprehensive research of renewable sources of energy commenced only after the energy crisis of the 1970s and became increasingly popular subsequently when *Renewable Energy: The Power to Choose*[33] was published in the mid-1980s. There are four generally defined distinct branches in the field of renewable energy studies: renewable energy policies analysis,[34] the relationship between renewable application and sustainable development,[35] the economic explanation for renewable energy, and an economic system design for renewable development.[36]

Notes

1　Oxford Dictionaries (2016).
2　UK Parliament (2016).
3　COP 21 (2015). Paris.
4　R. Mamberger (2003). *Energy Policy: Historical Overview, Conceptual Framework, and Continuing Issues*. New York: The Library of Congress.
5　Ibid.
6　L. M. Ross (2013). *How the 1973 Oil Embargo Saved the Planet*. Accessed May 2016 from www.foreignaffairs.com/articles/north-america/2013-10-15/how-1973-oil-embargo-saved-planet
7　B. Wang, J. Li and H. Wu (2014). Review and assessment of Chinese energy policy since the reform and opening up. *Emerging Markets Financial Trade*, 50, 143–158.
8　M. Kanellakis, G. Martinopoulos and T. Zachariadis (2013). European energy policy: A review. *Energy Policy*, 62, 1020–1030.
9　S. Carley (2011). The era of state energy policy innovation: A review of policy instruments. *Review Policy Research*, 28, 265–294.

10 H. C. Ong, T. M. I. Mahlia and H. H. Masjuki (2012). A review on energy pattern and policy for transportation sector in Malaysia. *Renewable Sustainable Energy Review*, 16, 532–542.

11 S. Z. Shirazi and S. M. Z. Shirazi (2012). Review of Spanish renewable energy policy to encourage investment in solar photovoltaic. *Journal of Renewable Sustainable Energy*, 4, 662–702.

12 K. H. Solangi, M. R. Islam, R. Saidur, N. Rahim and H. Fayaz (2011). A review on global solar energy policy. *Renewable Sustainable Energy Review*, 15, 2149–2163.

13 R. Saidur, M. R. Islam, N. Rahim and K. H. Solangi (2010). A review on global wind energy policy. *Renewable Sustainable Energy Review*, 14, 1744–1762.

14 H. Meyar-Naimi and S. Vaez-Zadeh (2012). Sustainable development based energy policy making frameworks, a critical review. *Energy Policy*, 43, 351–361.

15 W. M. Chen, H. Kim and H. Yamaguchi (2014). Renewable energy in eastern Asia: Renewable energy policy review and comparative SWOT analysis for promoting renewable energy in Japan, South Korea, and Taiwan. *Energy Policy*, 74, 319–329.

16 J. M. Blair (1976). *The Control of Oil*. New York: Pantheon Books.

17 E. J. Balleisen (2010). *Government and Markets: Toward a New Theory of Regulation*. Cambridge, MA: Harvard University Press.

18 W. H. Willam (2002). Energy modelling for policy studies. *Operations Research*, 50, 89–95.

19 A. B. Lovins (1977). *Soft Energy Path: Toward a Durable Peace*. San Francisco, CA: Friends of the Earth International/Ballinger Publishing Company.

20 J. MacKenzie (1977). Review of the nuclear power controversy by Arthur W. Murphy. *Quarterly Review of Biology*, 52, 467–468.

21 D. W. Fischer and E. Berglund (1994). The greening of Swedish energy policy: A critique. *Futures*, 26, 305–322.

22 IEA (International Energy Agency). *Energy Policies of IEA Countries: Sweden 2013 Review*. Paris: OECD Publications.

23 S. Pacala and R. Socolow (2004). Stabilization wedges: Solving the climate problem for the next 50 years with current technologies. *Science*, 305, 968–972.

24 IEA (International Energy Agency). *Energy Technology Perspectives: Scenarios & Strategies to 2050*. Accessed January 2015 from www.iea.org/textbase/npsum/etp.pdf

25 *Climate Solutions: The WWF Vision for 2050*. Accessed January 2015 from http://d2ouvy-59p0dg6k.cloudfront.net/downloads/climatesolutionweb.pdf

26 E. R. Berndt and D. O. Wood (1975). Technology, prices, and the derived demand for energy. *Review of Economics and Statistics*, 57, 259–268.

27 J. Griffin and P. Gregory (1978). An inter-country translog model of energy substitution responses. *American Economic Review*, 66, 845–857.

28 L. R. Christensen, D. W. Jorgenson and L. J. Lau (1973). Transcendental logarithmic production frontier. *The Review of Economics and Statistics*, 55, 28–45.

29 C. S. Craig and J. M. McCann (1978). Assessing communication effects on energy conservation. *Journal of Consumer Research*, 5, 82–88.

30 H. Hotelling (1931). The economics of exhaustible resources. *Journal of Political Economy*, 39, 137–175.

31 J. L. Sundquist (1969). *Making Federalism Work: A Study of Program Coordination at the Community Level*. Washington, DC: The Brookings Institution.

32 A. R. Light (1976). Federalism and the energy crisis: A view from the states. *Publius*, 6, 81–96.

33 D. Deudney and F. Christopher (1983). *Renewable Energy: The Power to Choose*. New York: W. W. Norton.

34 K. Kaygusuz (2003). Energy policy and climate change in Turkey. *Energy Conversion and Management*, 44, 1671–1688.

35 K. Kaygusuz and A. Kaygusuz (2002). Renewable energy and sustainable development in Turkey. *Renewable Energy*, 25, 431–453.

36 C. Mitchell and P. Connor (2004). Renewable energy policy in the UK 1990–2003. *Energy Policy*, 32, 1935–1947.

14 Measures, practices and tools

Utilising best practices to reduce air pollution and the carbon footprint worldwide

Economic, environmental, social and energy objectives may be achieved by making physical changes not just to the demand for goods and services but also to the systems providing the same.

These physical changes are brought about by people changing consumption patterns, by using technologies differently, and also by developing entirely new technologies.

This proposal aims at further work that will help collate measures for reducing greenhouse gas emissions and associated air pollutants all over the world and to identify the best practices for introducing these measures. Here Europe is being taken as the main example, with the focus being on emissions from energy supply and consumption. For example:

- Germany aspires to retrofit energy efficiency to its entire housing stock within the next two decades.
- Scandinavian nations have put in place an infrastructure that utilises a mix of a multi-fuelled district heating system with CHP and electric heat pumps.
- Generally European public transport usage is higher than other regions.
- Germany utilises more wind-generated energy than the UK in spite of having poorer wind resources.

The question remains – why can't these policies be applied universally? I undertook an analysis by compiling a comprehensive list of measures across all relevant sectors and collated best practice across Europe so as to discover the countries that have introduced the best measures while using different instruments. Then I estimated the effects of applying these best practices across Europe using different models. The resultant work is useful for the whole world, as other countries may learn to benefit from the best actual experiences in implementing low emission methods from the countries that have actually enforced these measures.

Of crucial importance to any integrated policy are multiple options such as better demand management, energy efficiency and the use of low-impact fuels. These are categorised in Table 14.1. The first four categories are colloquially referred to as 'Non-End of Pipe' (NEOP) options.

Generally the overarching emphasis has always been on EOP options for the control of non-GHGs. This is partly because significant levels of reduction in some pollutants such as SO_2 could be efficiently achieved at low cost with EOP technologies applied through relatively simple regulations such as the Large Combustion Plant Directive, without requiring whole-scale changes to the energy economy. Although,

Table 14.1 Emission control option categories

	Category	Examples
NEOP	Behavioural change	Economical vehicles, lower speeds
NEOP	Demand management	Efficient insulation for buildings, lower energy-consuming appliances, public transport demand
NEOP	Improved energy conversion	Utilising condensing boilers, CHP, combined cycle gas turbines
NEOP	Fuel switching	Switching from high-refining and high-pollution fuels such as coal and oil to gas and renewables
EOP	End-of-pipe	Flue gas desulphurisation, carbon sequestration catalytic converters (especially in newly manufactured vehicles along with carbon sequestration)

Note: NEOP – Non-End of Pipe; EOP – End of Pipe.

generally speaking, global warming and greenhouse gases have not been regulated by EU legislation, the EOP options' cost-effective limits have been reached, even as concern about global warming has correspondingly increased. A measure of political commitment to controlling it has followed both within individual member states and the EU as a whole; there has also been a greater emphasis on developing integrated policies that address the multiple environment problems of global warming and air pollution. This is chiefly because measures to control greenhouse gases tend to also reduce air pollutants, and thereby strict targets for both can be achieved at a lower total cost combined rather than separately.

Furthermore, EOP-abatement technologies typically decrease energy efficiency and some produce wastes, and decreasing energy efficiency usually causes a corresponding increase in CO_2 emissions.

For instance, flue gas desulphurisation may decrease the efficiency of electricity generation by approximately 5 per cent and also require limestone inputs and produce waste gypsum; carbon sequestration by pumping CO_2 into depleted reservoirs may also decrease energy efficiency substantially by around 10–35 per cent and hence increase primary CO_2 production and emission.

In addition to reducing greenhouse gas emissions, NEOP options generally decrease the emissions of air pollutants such as SO_2 and NO_x as well, because fossil fuel combustion is markedly reduced. NEOP options facilitate greater emission reduction than is possible with simple EOP measures alone, and the total combined cost of meeting greenhouse gases and air pollutant targets is generally less than in circumstances that do not include the extensive use of NEOP.

15 Electricity markets

All elements of an electricity system work simultaneously together in a coordinated way. Any decisions and/or actions made by one member will simultaneously affect all the members that are contributing to and make up the whole network.

Figure 15.1 provides a snapshot of the overall electricity market structure. This entire system consists of a large assortment of mutually interlinked generation, transmission and storage assets. These assets are partly complementary (all power plants must be connected to transmission lines, otherwise they cannot power anything other than their own selves) and in part are substitutes (a power plant supplying electricity locally may well be replaced by a transmission line that brings electricity from another plant).[1] As a result it is both challenging and difficult to administer such a system. For example, adding even a single transmission line may well result in the overloading of other power lines, causing them to trip and in effect lead to potential power outage in the entire area that is being supplied by the grid. Furthermore, new power plants just might require network extensions hundreds of kilometres away.[2] This is why all power systems demand proper administration and coordination along with well-structured policy frameworks and the investments necessary for their smooth functionality. Once these core requisites are in position, the electricity system would be able to guarantee energy efficiency, increase reliability security, and also reduce congestion in the system.

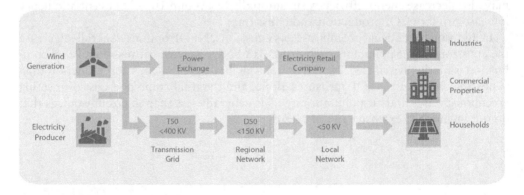

Figure 15.1 Overview of the electricity market structure.

15.1 Prototypes: electricity business and markets

The whole electricity market consists of myriad submarkets that are closely interrelated even as they operate in diverse ways and serve different purposes.

Electricity markets can be divided into two core groups, depending on where the actual trade takes place, namely the wholesale market and the retail market. The wholesale market is where most of the electricity is sold and purchased between generators, suppliers, non-physical traders and large end consumers.[3] Such trading aims to connect demand and supply at practical prices, and also strives to ensure effective management of both financial and physical risks. On the other hand, the retail market is where electricity is eventually sold to the small end consumer.[4]

Electricity submarkets consist of the following:

- forward markets: suppliers and customers can enter into contracts for electricity directly with generators to sell/buy electricity at a future date at a price agreed today;
- future markets: the contracts are traded on a future exchange;
- spot market: electricity will be traded and delivered only the next day. In the day-ahead market, power is traded during one day for delivery during the next day (e.g. NorthPool Spot);
- intraday market: plays the role of a balancing market to the day-ahead market;
- balancing and reserve market: is responsible for the balance between demand and supply in the system;
- congestion revenue rights, financial transmission rights and transmission congestion contracts: these financial instruments form separate electricity submarkets and are used to manage variability in congestion costs.

Therefore, it is reasonable to deduce that in the electricity markets (irrespective of type) competitive trading plays a vital function, as it helps to maintain markets' safety and transparency, as well as liquidity. Thus, it would not be wrong to state that energy trading is the core of the EU's (greatly) liberalised energy markets.[5]

15.2 Key market players

A number of players have an impact on the electricity markets:

- *Legislators and regulators* are responsible for drafting the legal framework.[6] They have the role of determining and/or approving the rules of the electricity market and investigating suspected cases of abuse of (market) power.[7]
- *Electricity producers* have by and large the twofold task of both generating electricity and then subsequently selling it in bulk in the market. However, the generation business lies at the heart of the electricity supply industry.[8]
- *Electricity suppliers* must be able to either buy or generate the electricity that they supply to other firms or large end consumers.[9]
- *Network companies* own their particular regional or local transmission network and therefore have the responsibility of transporting the electricity from the producer to the end consumers.[10] Network companies have to ensure reliability and efficiency of their electricity transportation/transmission network.
- *Power Exchanges (PXs)* have the responsibility of handling both 'day ahead and intraday' trading through implicit allocation of transmission capacity available

via Transmission System Operators (TSOs).[11] There are a large number of PXs spread throughout the European Union; in fact some countries have two or more exchanges. These PXs, working simultaneously with TSOs as well as national regulators, are key contributors to European energy goals.

- *Electricity retail companies* buy electricity either through the PXs or alternatively directly from the producer to sell it to end consumers.[12]
- *TSOs* meanwhile usually have a complete monopoly over transmission grid operations in a specific geographical area, and are therefore directly responsible for making sure that there is a constant balance of actual real time supply and demand[13] in their respective areas. Additionally they also coordinate at each border to allocate the transmission capacity of interconnectors for interzonal forward contracts that are directly agreed beforehand between the major market participants.[14]
- *Transmission firms* possess their own transmission system assets such as transformers, lines, cables and reactive compensation devices.[15]
- *Distribution System Operators (DSOs)* also own and run a very specific local 'low or medium voltage' network in a specific area and are also responsible for the distribution of electricity from producer to end user in that particular area.[16]
- A *Nominated Electricity Market Operator (NEMO)* has the responsibility of performing a single-day-ahead and/or single intraday MC. However, in line with new EC Regulation 2015/1222, there is a requirement that each and every Member State (MS) that is electrically connected to a bidding zone in another MS has to ensure that at least one (or more) NEMOs have been designed. The NEMO will also function as a market coupling operator. End consumers usually try to obtain security of supply at lower costs.[17]

15.3 A case study: European electricity markets – liberalisation and integration process

Among the nations that comprise the European Union, electricity networks as a rule tend to be regional monopolies;[18] these monopolies are encouraged by the Member States to reduce costs while improving quality, at the same time as also investing in transmission line expansion and technology innovation, along with adequate security of supply. Because of such rules, different regulatory arrangements have been developed in most EU countries. In an attempt to detach such natural monopolies from the competitive part of the sector, European legislation restricts joint control over both generation and transmission assets (a process referred to as 'unbundling').[19]

Transparent, reasonable and universal non-discriminatory access to the transmission network is of fundamental importance for a highly competitive wholesale electricity market to function properly.[20] Nevertheless, collusive behaviour is frequently observed in many such markets, and this attribute is highly noticeable in European wholesale markets as well.

Initially European electricity markets were basically organised as 'vertically integrated monopolies' and consequently were also managed the same way on a country to country basis. Due to this, all cross-border transactions took place within a cooperative structure among different national utilities, thereby concentrating more on system security as well as the efficient use of generation resources rather than concentrating on a purely commercial objective.[21] Consequently, national-level electricity markets in Europe have been exchanging electricity on a bilateral basis, i.e. only the lone buyer and seller are involved in the trade of electricity.

Nonetheless, such vertically integrated monopolies have been widely castigated vis-à-vis the remarkable liberalisation successes that have been a key characteristic of many other network industries. Many vertically integrated utilities have since been widely separated and unbundled, even as different entry barriers to both generation and supply are being eliminated so as to create a competitive environment.[22] Additionally, as of recent times bilateral trading is widely considered to be inefficient, and therefore competitive electricity pools have been formed so as to ensure that electricity trading is conducted in a centralised way and involves all producers and consumers alike.[23]

The various measures have been instrumental to the overall success of electricity industry reform throughout Europe. Most important have been two major directives – Directive 96/92/EC and Directive 2003/54/EC – and the comprehensive Third Legislative Package – Directive 2009/72/EC, Regulation 713/2009/EC and Regulation 714/2009/EC – that were promulgated to not only dismantle vertically integrated monopolies but also liberalise the production and supply of electricity as well.[24] They are also supposed to guarantee non-discriminatory network access to third parties while placing wider emphasis on both cross-border interconnections as well as the need to mitigate barriers to cross-border trade.[25]

Here is a brief overview of the temporal and geographic extent of European electricity markets:

- By 1921 the first network extensions for electricity transportation had been initiated.
- By 1921–1922 the project to liberalise and integrate a European energy market had been initiated.
- In 1925 the first supervisory body for transmission lines started.
- Between 1939 and 1949 the electricity sector in most European countries became owned and largely controlled by their governments.
- In 1957 the Treaty of Rome planned liberalisation for all commodity markets including the energy market.
- In 1959 the exchange of electricity became liberalised allowing national energy companies to more flexibly engage in corresponding transactions.
- In 1987 the Single European Act was a step towards abolishing state-owned national monopolies.
- In 1990 the liberalisation process of European electricity wholesale markets was initiated.
- In 1992 the Maastricht Treaty further strengthened European ambition to create free cross-border trade in electricity.
- 1996 saw the First Electricity Directive 1996/92/EC on electricity liberalisation.
- In 2000 the Florence Forum reached early agreements concerning market-based mechanisms for congestion management.
- 2003 saw the Second Electricity Directive 2003/5-4/EC, concerning common rules of the internal electricity market.
- In 2004 the EC set out its strategy on electricity market integration.
- In 2005 a second internal market package for electricity was adopted.
- In 2006 electricity regional initiatives were launched.
- In 2009 the third internal market package for electricity was adopted.
- In 2011 the EU adopted new stringent rules on wholesale electricity trading (Regulation 1227/2011). The European Council also set the target of 2014 for the completion of the internal electricity market.

- 2014 saw multiregional price coupling in north-western Europe (NWE).
- In 2015 a new EU regulation, (EC) 2015/1222, established a guideline on capacity allocation and congestion management.

15.4 Electricity market coupling

a) Principle and technique of market coupling

Market coupling (MC) is a means of joining and eventually integrating various energy markets into one cross-border market.[26] The TSOs are responsible for capacity trading, while PXs are responsible for the electricity itself. Such separate capacity trading is termed explicit auction. MC gives an opportunity to couple these separate markets together. It may be described as a mechanism that takes cross-border transmission capacities (TCs) and by means of implicit auctioning enables trade between two (or more) PXs.

As the requests stem from different exchanges, each of which represents a different network location[27] in such a coupled market, demand and supply requests in one market are no longer confined to the local territorial scope of that particular market.[28] Therefore, prices and TC are shared between two individual market regions that ultimately help to guarantee effective capacity utilisation.

TSOs also perform the function of providing information about available TCs along with the security of the grid between different market areas. This is why TSOs use a central coupling algorithm that provides information about prices and power flows. PXs operate specific price zones that divide the energy market (see Table 15.1 for the EU-28 PXs). These price zones have to be interlinked, and to do that well TSOs have to provide transfer capacities to the integrated PXs. This is why energy trading is subject to the capacity constraints set by the TSOs, and that may in turn limit electricity flows between different markets.[29]

The following subsection provides a snapshot of different ways of how to implement MC.

b) Methods of market coupling implementation

Typically there are many techniques for implementing MC; the most common, price-based MC (e.g. flow-based MC), means a high level of market integration[30] and is preferred in the EU. This method sets prices and flows using a coupling algorithm in a two-step process: first, the national TC is optimised (cross-border trading is not considered yet); and second, optimisation is implemented by means of consideration of international flows. This results in the determination of comprehensive market outcomes, inclusive of both prices and optimal power flows in each national market.

There may be a distinction between 'tight' and 'loose' volume-based MC. However, that depends on the extent to which the coupling algorithm is built into the set of information and matching rules characterising the domestic market outcomes.[31] For a detailed analysis of these MC approaches, see Tennet (2010).[32]

15.5 Market coupling: advantages and problems

The concept of market coupling has many advantages in the coupled regions; however, there are also disadvantages and issues that cast doubts on this mechanism.

Table 15.1 Electricity power exchanges in EU-28 and Norway

Country	Power exchange	Country	Power exchange
Austria	EXAA EPEX SPOT	Latvia	Nord Pool Spot
Belgium	Belpex (Part of APX)	Lithuania	Nord Pool Spot
Bulgaria	IBEX	Luxembourg	–
Croatia	CROPEX	Malta	–
Cyprus	–	Netherlands	APX ICE ENDEX
Czech Republic	OTE PXE	Norway	Nord Pool Spot
Denmark	Nord Pool Spot	Poland	POLPX PXE
Estonia	Nord Pool Spot	Portugal	OMIE
Finland	Nord Pool Spot	Romania	OPCOM PXE
France	EPEX SPOT	Slovakia	PXE
Germany	EPEX SPOT – EEX	Slovenia	BSP Regional Energy Exchange LL C
Greece	LAGIE	Spain	OMIE
Hungary	HUPX	Sweden	Nord Pool Spot
Ireland	SEMO	United Kingdom	APX N2EX – Nord Pool Spot ICE
Italy	IPEX – GME		

Sources: Baritaud and Volk, 2014; Imran and Kockar, 2014; EC, 2014.[33,34,35]

As is evident, MC seems beneficial, but 'its implementation requires the synchronisation of a large group of stakeholders',[36] not TSOs and PXs alone. There are different organisations, each with various preferences, and this has a tendency to increase uncertainty in energy trading – for instance, organisations (among them TSOs) interested in transmission investments, or for that matter even generation companies that are concerned about the location of new investments.

All the factors that are mentioned in this section create doubts over the success of the MC as well as the advantages that it may be able to (potentially) provide. Despite this, I agree with Zachmann[37] that MC is a system that 'in principle would ensure that price differentials between countries only arise when no additional transmission capacity may be available'.

15.6 Market coupling in the EU

MC is one of the main instruments that is expected to help establish a single main European electricity market by interconnecting national markets and by eliminating both physical and economic barriers. Such a single market is liable to guarantee a secure electricity supply at highly competitive prices all over Europe.

However, it is pertinent to note that the MC is more than just an agreement between PXs and TSOs. There are many more parties concerned in this particular approach, such as national regulators and governments, energy companies and producers, and also European Union authorities.[38]

This is why during the past ten years several MCs have taken place among neighbouring European markets, thus effectively enabling an implicit cross-border trade in electricity.[39] In fact, any markets on a regional level have already been coupled, and this in turn may be seen as a step-wise process towards the creation of a single EU-wide electricity market.[40] Table 15.2 provides a historical overview of the key MC initiatives all across Europe.

Therefore, MC can be seen as one of the most influential mechanisms leading to the integration of cross-border power trading arrangements all across Europe. Moreover, the possibility of reserving TC through auctions has allowed for better cross-border trade,[41] leading to higher electricity price convergence all across Europe.

As a consequence of effective trading, MC has increased wholesale electricity price convergence across borders; however, 'coupled markets do not necessarily lead to permanent price convergence in electricity prices across the coupled area'.[42] This shows that more research needs to be undertaken in order to improve convergence in prices.

The most crucial measure of European market integration took place on 4 February 2014 when Price-Based Coupling in NWE went live.[43] It was the very first project to use the pan-European Price Coupling of Regions (PCR) solution for calculating prices and flows – the starting point for all other regions to join.[44] The next section is concerned with the NWE region and an analysis of day-ahead prices in coupled electricity markets.

15.7 Multiregional market coupling

The core purpose of the NWE project was the implementation of a day-ahead electricity MC in the whole region based purely on the 'price coupling principle', using a single algorithm that would calculate simultaneously the market prices, net positions and flows on interconnectors between market areas.[45] It is imperative that this single algorithm meets all the TSOs' requirements in terms of efficient allocation of production, consumption and capacity and is also approved by every Member State.[46]

At least four PXs along with 13 TSOs have been involved in the NWE project. It coupled the day-ahead markets all across central western Europe (CWE), the Nordic countries, Great Britain and the Baltic countries, covering around 75 per cent of Europe's electricity demand.[47]

Since the launch of the NWE initiative, two extensions of the PCR-coupled area have already taken place: in February 2015, Italy coupled with France, Austria and Slovenia,[48] and in May 2014, Spain and Portugal joined. Hence the entire coupled area now covers 19 countries and is called Multiregional Coupling. The NWE project is responsible for about 85 per cent of European power consumption.

Table 15.2 Historical overview of the main MC initiatives across Europe

Region	Year when started	Countries or regions
Scandinavia	1993	Scandinavian market
Northern Region (Nord Pool)	1996 1998 2000 2010 2013	Sweden, Norway Denmark Finland Estonia Poland, Latvia
Northwest Europe	2006	France, Belgium, Netherlands
Spain–Portugal	2007	Spain, Portugal
Central Western Europe (CWE)	2007	Denmark, Belgium, France, Germany, Luxembourg, the Netherlands
	2008 2009	CWE TSOs Germany, Denmark
Central Eastern Europe (CEE)	2008	CEE countries
Poland–Sweden	2010	Poland, Sweden
CWE–Nordic	2010	Both regions are linked between: • the Netherlands and Norway • Germany and Denmark • Germany and Sweden
Slovenia–Italy	2011	Slovenia, Italy
CWE–Nordic–UK	2012	CWE–Nordic–UK
Central and Eastern Europe (CEE)	2012	Czech Republic, Slovakia, Hungary, Romania
South-Western Europe (SWE)	2014	France, Spain, Portugal
North-Western Europe (NWE)	2014	Nordic countries (including Baltic States, Poland and Sweden), Great Britain and CWE. Later SWE joined.
Italy–France–Austria–Slovenia	2015	Italy, France, Austria, Slovenia

PCR is based on three main principles: robust operation, a single algorithm, and lastly individual PX accountability.[49] The NWE project is the very first initiative to use the pan-European PCR solution for the calculation of prices and flows. It is also the starting point for all other regions to join.[50]

Lastly, it is necessary to refer to a gap in the studies covering multiregional MC. The annual reports ACER and CEER (2014) on key developments in EU electricity wholesale markets, including a comprehensive assessment of the level of market integration, have been published. However, their most recent report covers data up until 2013 only and they analyse only interregional price convergence, without considering the multiregional case. The EC (2014) also published a report in which it considered various recent developments in the wholesale electricity markets. However, there was no mention of the NWE initiative project. This is why no recent studies have been found that would analyse a multiregional MC approach based on a project implemented in reality.

15.8 Price coupling of regions within the EU

The first Memorandum of Understanding (MoU) on Price Coupling of Regions (PCR) was inked in June 2009 between EPEX OMIE and Nord Pool. In March 2010 a second one included GME and APX. In 2011, the same Price Coupling System was also introduced in CWE market coupling as well.

EPEX, Belpex, APX, GME, Nord Pool and OMIE in June 2012 signed two agreements on PCR: the cooperation agreement and the co-ownership agreement. In April 2013, OTE adhered to the PCR cooperation agreement.[51,52,53]

In February 2014, PCR was launched for Finland, Denmark, Belgium, France, Estonia, Germany, Latvia, Austria, the UK, Lithuania, Luxembourg, the Netherlands, Norway, Poland (via the SwePol Link) and Sweden. The SWE region (France–Spain–Portugal) also operated with the same algorithm, but the French/Spanish cross-border capacities allocation remained the same.[54,55] In May 2014, SWE was successfully introduced in the PCR when the cross-border capacities between France and Spain were for the first time allocated via implicit auctions.[56]

a) Overview of existing studies

Directives as well as regulations nudging the EU towards a single European electricity market have grown over the past decade. I concentrate on the research studies that assess the level of market integration based on price analysis.

Bower[57] and Boisseleau[58] were among the very first researchers to study European power market integration. Boisseleau analysed the power exchanges in Germany and the Netherlands for the year 2001. The study concluded that there was low integration between power exchanges and price differences even between the two German exchanges themselves and better integration on long-term contracts than short-term ones.

Bosco *et al.* (2007)[59] confirm and further reinforce the conclusions of Bower (2002) and Boisseleau (2002). They studied at length the weekly medians of the hourly time series of six wholesale markets from 1999 (for already existing exchanges) to 2007: APX (the Netherlands), EEX (Germany), EXAA (Austria), Nord Pool (Scandinavia), Omel (Spain) and Powernext (France). Weekly medians have the advantage of eliminating daily as well as weekly seasonality. They have found better cointegration for a group of core countries (the Netherlands, Germany, France and Austria). They have also concluded that there is strong integration between Germany and France.

Armstrong and Galli (2005)[60] used an exploratory data approach to comprehensively analyse the hourly day-ahead differences between Germany, France, the Netherlands and Spain. Their findings show that markets may be converging but convergence has not been fully achieved as yet. The findings also indicate that variability is decreasing over time. However, they emphasise both a lack of transparency as well as transmission constraints between these two nations. On the other hand, Zachmann (2008)[61] clearly distinguishes the level of market integration ('the static degree to which the single European market is attained') as well as price convergence ('the dynamic measure for the development of prices toward a single overall European price') in his analysis. He applies a 'Principal Component Analysis' (PCA) to the hourly wholesale prices of ten countries (Austria, Czech Republic, the Netherlands, Denmark (east and west), Germany, France, Spain, Poland, Sweden and United Kingdom) between 2002 and 2006. The conclusion of the study was that there is price convergence for pairs of countries, especially in directly connected countries.

A problem that emerged during the course of the various studies was that in order to analyse and synthesise all these results there was no specific time frame, while the research covers different data periods, countries and types of contracts. However, the authors worldwide concluded that there is relative integration for core markets, with higher convergence in neighbouring countries, but still a low integration for Europe as a whole.

b) Advantages of market coupling

The greatest benefit of MC lies in its ability to enhance competition. It has the dual effect of decreasing prices while also limiting the abuse of market power.

Welfare advantages

The following example helps us study the welfare gains of coupling three different power exchanges. Figure 15.2 shows the corresponding supply and demand curves.

The clearing prices and volumes correspond to the intersections of the supply and demand curves. Where coupled exchanges are concerned, the bids of the three exchanges are ranked from high to low and the corresponding offers from low to high. Therefore, the supply and demand curves are an aggregate of the curves obtained for the exchanges without coupling.

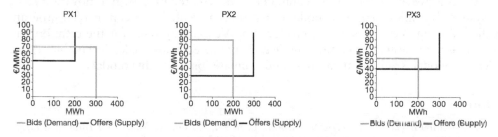

Figure 15.2 Supply and demand curves for three power exchanges.

Market power

Using market power consists of withholding generation capacities so as to artificially increase the equilibrium price or to foist prices that don't usually reflect the generation cost in coordination with the other generators. This is referred to as 'collusive behaviour'. By increasing the overall number of market players, MC decreases the risk of abusing market power. In fact, the smaller the share of a generator, the less likely that price could be influenced, and correspondingly the more players there are, the less likely there will be an agreement to manipulate the market. Furthermore, even if a generator does withhold capacity (in a bid to increase price), the cumulative effect on price is highly diluted due to the increased number of players.

c) Price correlation

The results underscore the existence of at least four regional markets with a high correlation coefficient: CWE (74 per cent of correlation on average within the region), central and eastern Europe (CEE)[62] (82 per cent), MIBEL (98 per cent) and Nord Pool, and also highlight the fact that CEE countries are pretty well integrated. Moreover, the region shows signs of growing convergence with the rest of Europe, especially CWE. As a matter of fact, 13 pairs of countries involving one CWE and one CEE country have a correlation coefficient higher than 0.77.

15.9 Conclusions

The overall analysis of prices clearly shows that there are several European regions where price correlation is (relatively) high. Moreover, the correlation between regions is also increasing, especially between CWE and CEE. The MC is often considered as an ongoing process that links several markets that always stay independent rather than as a single market. Only market splitting or nodal pricing could (potentially) enable the creation of a single electricity market. This would result in an increased cooperation between national markets but never in the creation of a single electricity market.[63]

The Energy Policy Simulator (EPS) is a system dynamic model that assesses the effects of 51 energy and environmental policies on a variety of metrics, including the emissions of 12 pollutants; cash flow changes for government, industry and consumers; the composition of the electricity generation fleet; the usage of various fuels; and the monetised social benefits from avoided public health impacts and climate damage.

EPS is a free and open-source computer model created by Energy Innovation LLC; a system dynamics computer model has also been created in a commercial program called Vensim. Vensim is a tool produced by Ventana Systems for the creation and simulation of system dynamics models. Ventana Systems makes available a free Vensim Model Reader that can read and simulate (but not edit) models.

Notes

1 G. Zachmann (2015). *Electricity without Borders: A Plan to Make the Internal Market Work*. Brussels: Bruegelblueprint. Accessed July 2015 from http://bruegel.org
2 Ibid.
3 EWEA (2012). *Creating the Internal Energy Market in Europe*. Accessed June 2015 from www.ewea.org/uploads/tx_err/Internal_energy_market.pdf
4 Ibid.

5 European Federation of Energy Traders (EFET) (2012). *Towards a Single European Energy Market: The Basics of Wholesale Energy Trading; The Role of EFET; Our Vision for Further Market Development.* Amsterdam: EFET. Accessed July 2015 from www.efet.org

6 Vattenfall (2011). *The Common European Energy Market: Electricity, Gas and Heat.* Stockholm: Vattenfall AB. Accessed August 2015 from https://corporate.vattenfall.com

7 D. Kirschen and G. Strbac (2004). *Fundamentals of Power System Economics.* Manchester: John Wiley & Sons. Accessed July 2015 from http://onlinelibrary.wiley.com

8 P. Mäntysaari (2015). *EU Electricity Trade Law: The Legal Tools of Electricity Producers in the Internal Electricity Market.* Switzerland: Springer International Publishing.

9 Ibid.

10 Vattenfall (2011). *The Common European Energy Market: Electricity, Gas and Heat.* Stockholm: Vattenfall AB. Accessed August 2015 from https://corporate.vattenfall.com

11 K. Imran and I. Kockar (2014). A technical comparison of wholesale electricity markets in North America and Europe. *Electric Power Systems Research*, 108, 59–67.

12 Vattenfall (2011). *The Common European Energy Market: Electricity, Gas and Heat.* Stockholm: Vattenfall AB. Accessed August 2015 from https://corporate.vattenfall.com

13 Ibid.

14 K. Imran and I. Kockar (2014). A technical comparison of wholesale electricity markets in North America and Europe. *Electric Power Systems Research*, 108, 59–67.

15 European Federation of Energy Traders (EFET) (2012). *Towards a Single European Energy Market: The Basics of Wholesale Energy Trading; The Role of EFET; Our Vision for Further Market Development.* Amsterdam: EFET. Accessed July 2015 from www.efet.org

16 Vattenfall (2011). *The Common European Energy Market: Electricity, Gas and Heat.* Stockholm: Vattenfall AB. Accessed August 2015 from https://corporate.vattenfall.com

17 P. Mäntysaari (2015). *EU Electricity Trade Law: The Legal Tools of Electricity Producers in the Internal Electricity Market.* Switzerland: Springer International Publishing.

18 G. Zachmann (2015). *Electricity without Borders: A Plan to Make the Internal Market Work.* Brussels: Bruegelblueprint. Accessed July 2015 from http://bruegel.org

19 Ibid.

20 European Federation of Energy Traders (EFET) (2012). *Towards a Single European Energy Market: The Basics of Wholesale Energy Trading; The Role of EFET; Our Vision for Further Market Development.* Amsterdam: EFET. Accessed July 2015 from www.efet.org

21 G. Squicciarini, G. Cervigni, D. Perekhodtsev and C. Poletti (2010). *The Integration of the European Electricity Markets at a Turning Point: From the Regional Model to the Third Legislative Package.* EUO Working Papers RsCAS 2010/56, Robert Schuman Centre for Advanced Studies Florence School of Regulation. Accessed February 2014 from http://fsr.eui.eu/Documents/WorkingPapers/Energy/2010/WP201056.pdf

22 L. Meeus, K. Purchala and R. Belmans (2005). Development of the internal electricity market in Europe. *The Electricity Journal*, 18 (6), 25–35.

23 D. Kirschen and G. Strbac (2004). *Fundamentals of Power System Economics.* Manchester: John Wiley & Sons. Accessed July 2015 from http://onlinelibrary.wiley.com

24 E. Pellini (2014). *Essays on European Electricity Market Integration.* PhD thesis, University of Surrey.

25 T. Jamasb and R. Nepal (2012). *Market Integration, Efficiency and Interconnectors: The Irish Single Electricity Market.* Edinburgh: Heriot-Watt University. Accessed August 2015 from http://papers.ssrn.com/sol5

26 Price Coupling of Regions (PCR) (2015a). *Euphemia Public Description – PCR Market Coupling Algorithm.* Accessed August 2015 from www.epexspot.com/document/35382/Euphemia%20Public%20Description%20-%20August%202016.pdf

27 L. Meeus, K. Purchala and R Belmans (2005). Development of the internal electricity market in Europe. *The Electricity Journal*, 18 (6), 25–35.

28 Price Coupling of Regions (PCR) (2015b). *Euphemia Public Description – PCR Market Coupling Algorithm.* Accessed August 2015 from www.epexspot.com/document/35382/Euphemia%20Public%20Description%20-%20August%202016.pdf

29 EWEA (2012). *Creating the Internal Energy Market in Europe.* Accessed June 2015 from www.ewea.org/uploads/tx_err/Internal_energy_market.pdf

30 P. Mäntysaari (2015). *EU Electricity Trade Law: The Legal Tools of Electricity Producers in the Internal Electricity Market.* Switzerland: Springer International Publishing.

31 G. Buglione, G. Cervigni, E. Fumagalli, E. Fumagalli and C. Poletti (2009). *Integrating European Electricity Markets, Research Report Series – iSSN 2036-1785*. IEFE Centre for Research on Energy and Environmental Economics and Policy. Accessed January 2014 from www.iefe.unibocconi.it/wps/allegatiCTP/RR_No_2_ERI_viola_1.pdf

32 Tennet (2010). *Market Integration: Coupling of the European Electricity Markets*. Amsterdam: Tennet TSO B.V. Accessed August 2015 from www.tennet.eu/nl/marktkoppeling

33 M. Baritaud and D. Volk (2014). *Seamless Power Markets – Regional Integration of Electricity Markets in IEA Member Countries*. France: International Energy Agency (IEA).

34 K. Imran and I. Kockar (2014). A technical comparison of wholesale electricity markets in North America and Europe. *Electric Power Systems Research*, 108, 59–67.

35 European Council (EC) (2014). *Country Profiles. Annex II*. Single Market Progress Report. Accessed August 2015 from https://ec.europa.eu/progress-report

36 Price Coupling of Regions (PCR) (2015b). *Euphemia Public Description – PCR Market Coupling Algorithm*. www.epexspot.com/document/35382/Euphemia%20Public %20Description %20-%20August%202016.pdf

37 G. Zachmann (2015). *Electricity without Borders: A Plan to Make the Internal Market Work*. Brussels: Bruegelblueprint. Accessed July 2015 from http://bruegel.org

38 Tennet (2010). *Market Integration: Coupling of the European Electricity Markets*. Amsterdam: Tennet TSO B.V. Accessed August 2015 from www.tennet.eu/nl/marktkoppeling

39 European Commission (EC) (2014). *EU Energy Markets in 2014*. Belgium: Publication Office of the European Union. Accessed August 2015 from http://ec.europa.eu/energy

40 V. Böckers and U. Heimeshoff (2014). The extent of European power markets. *Energy Economics*, 46, 102–111.

41 G. Zachmann (2015). *Electricity without Borders: A Plan to Make the Internal Market Work*. Brussels: Bruegelblueprint. Accessed July 2015 from http://bruegel.org

42 European Council (EC) (2014). *Country Profiles. Annex II*. Single Market Progress Report. Accessed August 2015 from https://ec.europa.eu/progress-report, p. 24.

43 EPEX SPOT (2014a). *Market Data: Day-Ahead Auction*. Accessed August 2015 from www.epexspot.com/en/market-data

44 Ibid.

45 Belpex (n.d.). *Belpex Market Results*. Accessed June 2015 from www.belpex.be

46 NordREG (2014). *Nordic Market Report 2014: Development in the Nordic Electricity Market*. Sweden: Nordic Market Regulator. Accessed August 2015 from www.nordic energyregulators.org

47 Nord Pool Spot n.d. (a). *North-Western European Price Coupling (NWE)*. Accessed August 2015 from www.nordpoolspot.com/How-does-it-work/Integrated-Europe/NWE

48 EPEX SPOT (n.d.). *Market Coupling – A Major Step Towards Market Integration*. Accessed August 2015 from www.epexspot.com/en/market-coupling

49 Nord Pool Spot n.d. (b). *Price Coupling of Regions*. Accessed August 2015 from www. nordpoolspot.com/How-does-it-work/Integrated-Europe/Price-coupling-of-regions

50 EPEX SPOT (2014c). *Market Coupling: A Major Step Towards Market Integration*. Accessed February 2016 from www.epexspot.com/en/market-coupling

51 EPEX SPOT (2014b). Accessed January 2014 from www.epexspot.com/en/company-info

52 EPEX SPOT (2014c). *Glossary*. Accessed February 2016 from www.epexspot.com/en/extras/glossary

53 EPEX SPOT (2014c). *Market Coupling: A Major Step Towards Market Integration*. Accessed February 2016 from www.epexspot.com/en/market-coupling

54 APX (2014a). *CWE Flow-Based Market Coupling*. Accessed February 2016 from www. apxgroup.com/services/research-projects/cwe-flow-based-market-coupling

55 APX (2014b). *North-Western European Power Markets Successfully Coupled*. Accessed February 2016 from www.apxgroup.com/press-releases/north-western-european-power-markets-successfully-coupled

56 Energinet (2014). *South-Western and North-Western Europe Day-Ahead Markets Successfully Coupled*. Accessed February 2016 from http://energinet.dk/Site CollectionDocuments/ Engelske%20dokumenter/El/SWE_full_coupling_go-live_confirmation.pdf

57 J. Bower (2002). *Seeking the Single European Electricity Market: Evidence from an Empirical Analysis of Wholesale Market Prices*. Oxford: Oxford Institute for Energy Studies.

58 F. Boisseleau (2002). *The Role of Electricity Trading and Power Exchanges for the Construction of a Common European Electricity Market*. IEEE.

59 B. Bosco, P. Lucia, M. Pelagatti and F. Baldi (2007). *A Robust Multivariate Long Run Analysis of European Electricity Prices*. Milan: Fondazione Enrico Mattei.

60 M. Armstrong and A. Galli (2005). *Are Day-Ahead Prices for Electricity Converging in Continental Europe? An Exploratory Data Approach*. Paris: CERNA.

61 G. Zachmann (2008). Electricity wholesale market prices in Europe: Convergence? *Energy Economics*, 30, 1659–1671.

62 Romania is excluded because the data was not available.

63 A. Ouriachi and C. Spataru (2015). Integrating regional electricity markets towards a single European market. *International Conference on the European Energy Market, EEM, August 2015*.

16 Policy instruments and marketing of renewables in regions worldwide

16.1 Introduction

Across the globe countries are developing their renewable energy plans for several reasons: to reduce carbon emissions, to lower the use of fossil fuels, and to provide more reliance and security. Given their higher costs compared to conventional forms of electricity production and their intermittent nature that impacts on system dispatchability, regulatory intervention has been necessary for them to grow. The challenge for governments and regulatory agencies consists in choosing the type of support mechanisms that should be implemented to ensure their development, especially considering they always clash with existing economic and industrial policies. Concerns vary from country to country, depending on the prevailing generation technology, its degree of sectorial maturity[1] and its dispatchability.[2] In general, discussions are around security of supply, the increased prices that may place tariff burdens on final consumers and the regulatory risks faced by investors. For example, schemes or changes in regulation put at stake the generator's capacity to meet expected rate of return, which results in increased financing costs. On the other hand, risk-mitigating schemes can reduce the levelised cost of electricity and consequently support costs to consumers.[3]

There are several direct and indirect methods of support in place. The latter include, among others, research and development funding, regulation facilitating grid access, and net metering.[4] On the other hand, direct methods include both price-based and quantity-based mechanisms. This chapter will look into the several direct regulatory support mechanisms in place for deployment of renewable energy (RE) in the EU and South America.

16.2 Current support mechanisms for deployment of RE

Governments have several choices concerning their support. They can either reduce the risk by removing the main barriers (e.g. administrative, grid connection), transfer the risks to public actors (e.g. soft loans) or implement direct financial support. Direct and de-risking mechanisms can be categorised into price-driven policies and quantity-driven policies.[5] For price-driven policies, the price of energy is fixed by the government and the quantity is then set by the market, while for quantity-driven policies the quantity is imposed by the policy and the market then sets a price.

There are several mechanisms currently available under each category (Table 16.1). Support schemes under price-based policies include: Feed-In-Tariffs (FIT), Feed-In-Premiums (FIP), Tax Incentives and Investment Incentives; while quantity-based policies include: Tradable Green Certificates (TGC) and competitive tenders. All of

Table 16.1 Overview of the existing support mechanisms

		Direct		Indirect
		Price-driven	*Quantity-driven*	
Regulatory	*Investment focused*	Investment incentives Fiscal incentives	Tendering	Environmental taxes
	Generation based	Feed-In-Tariffs Rate-based incentives	Tendering quota obligations (RPS) based on TGCs	
Voluntary	*Investment focused*	Shareholder programmes Contribution programmes		Voluntary agreements
	Generation based	Green tariffs		

which have impacted on both the increase of the RES share in the final consumption and energy efficiency worldwide. The current common practice is for countries to choose a combination of these mechanisms. For example, in the EU the majority of its member states implemented FIT combined with either FIP or tenders. In contrast, in South America most countries have opted for long-term tenders, and price-based mechanisms are disappearing from regulation in the region.[6]

An overview of the existing support mechanisms is provided in Table 16.1.[7]

a) Price-driven mechanisms

Feed-In-Tariffs (FIT)

For FIT, the prices are imposed by the government (a price per MWh of RE to all the producers) and set depending on the technology used; they need to be developed in order to make these technologies economically interesting for producers. To ensure stability for investors, these prices need to be stable for a certain period – thus with long-term certainty the investment risk is reduced. Unlike quantity-based mechanisms such as quotas, FIT does not expose RE developers to risk in the certificates market, since there is no competition among RE providers. The prices can be fixed for the whole period of the support mechanism (generally ten years) or decreased gradually over time in case of an important learning rate or a lower generation cost than expected. It is thus important that these prices are reviewed regularly in order to be cost-efficient and to avoid windfall profits for the producers.

FIT has been used by several countries. It provides the required support for RES technologies that have moved beyond the R&D phase and still lack market maturity, such as solar PV.[8] The main disadvantage of FIT lies in the challenge of setting the right remuneration to stimulate investors, while avoiding conditions of over-support and overpayment, especially for technologies that have similar development costs for all developers.[9] Moreover, guaranteed revenue linked only to production eliminates the need for producers to react to market price signals, which puts at stake the control and reduction of systems imbalances.[10] Finally, there are risks to be taken into consideration because of changes in government and political preferences. For instance, in Spain measures have been taken to reduce retroactively the remuneration initially agreed on the first regulatory instruments implementing FIT in the country.

Feed-In-Premiums (FIP)

FIP are payments ensured to producers on top of the market price. These are generally implemented in parallel to FIT. Producers can choose between the two mechanisms. Producers are directly linked to the electricity market and can thus adjust their production to the demand. This is especially true for RES technologies such as biomass plants because of their dispatchable character with the possibility of increasing production in response to higher prices. The downside is that it can create incentives for gaming via predictions and result in inefficient dispatch, thus contributing to potentially higher prices and premiums. Finally, because of the risks associated with FIP being granted on top of market prices, new generators face greater barriers than vertically integrated companies, which may result in a competitive advantage to the latter and in market power problems.

Tax and investment incentives

Tax incentives often represent an accelerated depreciation schedule. There can also be tax exemptions for producers. This depends on the installed capacity and total renewable energy production. It thus represents an avoided cost compared to conventional electricity. There are two main types of investment incentives: investment subsidies and soft loans. For investment subsidies, an incentive is given to future producers, which depends on their installed capacity. For soft loans, the government loans money to future producers with attractive interest rates and longer repayment periods. Finally, restrictions on import duty are used to support local developers of RE technologies.[11] Such incentives are only advantageous from an equity perspective. Thus, they have proved to be more attractive for national producers who have large revenue flows and enough to turn the tax credits and other incentives into cash for their own operation.

b) Quantity-driven mechanisms

Quota obligations with TGCs

Quota obligations are also known as renewable obligations (RO). These are minimum shares of renewable energy imposed by the government on different 'actors': suppliers, consumers or producers. Penalties are imposed when they can't meet their obligation, and are redistributed proportionally to the actors that have met their obligations, according to the share of RE they have reached. In parallel, certificates called TGCs are distributed to the producers. These can then be sold to the consumers. The producer's assurance comes from the money gained as a result of the electricity consumed and from the sale of its certificates. Depending on the technology used, the number of certificates allocated per MWh varies. For example, for offshore wind there are two certificates per MWh.[12] The price of TGCs is automatically adjusted by considering the generation costs of the technology. This results in a higher uncertainty for investors, which is generally transferred to consumers through higher prices. However, the main advantage is that transnational and inter-state efficiencies can be better supported through trading of certificates across different geographies.

Tendering system

In this case, the government asks producers to bid in order to find the least costly producer. The winner will receive a long-term contract with favourable conditions such as inclusion of subsidies. There are several different design options that will determine who has to pay for the related costs in supporting RE, who is responsible for the demand forecasting and auctioning process, who can take part in the procedure and under what conditions, as well as the contract conditions and whether it is centralised or decentralised.[13] The main point that distinguishes RE auctions from other auctions is in the methodology used to administratively determine the firm capacity of the RE projects. This mechanism shares a number of advantages with the FIT, such as reduced risk for producers; and instead of the regulator identifying the costs associated with RE technology, it is the market that does so through bidding procedures.[14]

16.3 Regional case studies

European Union (EU)

Renewables and targets

The EU is composed of countries with very different characteristics. Their total electricity mix varies significantly (Figure 16.1). Also their administrative framework and their

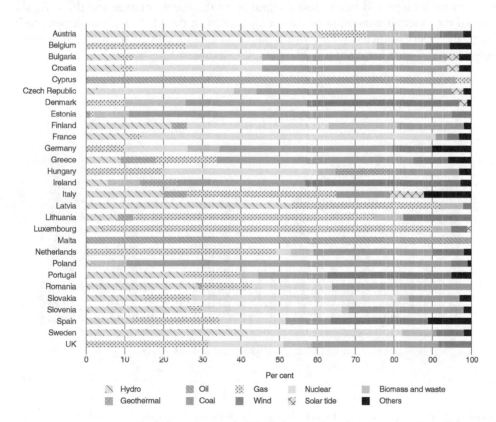

Figure 16.1 EU member states' energy mix (shares in gross inland consumption). Data source: 15

grid connection systems are different, which makes the design of a single European policy system nearly impossible.

The European Commission (EC) has set for each member state a national target for renewable energy (Figure 16.2) and one for energy efficiency. In 2013, the majority of member states reached more than half of their established targets for 2020. In some cases, countries such as Sweden surpassed their original commitments (Figure 16.2).

Directives

The European Parliament has implemented the following main directives to support RE: Directive 2001/77/EC (2001) for the promotion of electricity produced from renewable energy sources, and Directive 2003/30/EC (2003) on the promotion of the use of renewable fuels for transport.

The regulatory approach to renewables in the EU is very closely connected to the Union's goals related to combating climate change and promoting sustainability. In 2007 an ambitious set of objectives related to energy and climate change was introduced. By 2020 EU member states are required to reduce carbon emissions by 20 per cent in relation to 1990 levels, increase the total share of renewable energy sources by 20 per cent and improve energy efficiency by 20 per cent.[16] These main EU goals were included in the so-called '2020 Strategy', aimed at achieving a more sustainable and inclusive growth for all member states. The strategy prioritised efficiency, empowerment of consumers, leadership in energy technology and innovation, integration of the energy market and the strengthening of the external dimension of the energy market in the EU.[17] To complement such policy, the Energy Efficiency Directive required member states to set national primary energy saving targets for 2020 to achieve end-use saving of 1.5 per cent.[18]

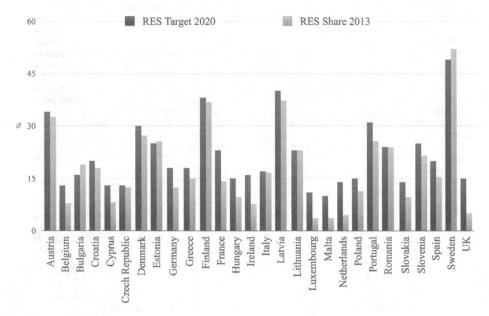

Figure 16.2 Target versus actual reached values in 2013 for electricity. [Data source: 19,20]

To meet even longer-term climate objectives, new targets were proposed under the EU 2030 Strategy: a 40 per cent cut in greenhouse gas emissions compared to 1990 levels, at least 27 per cent share of renewable energy consumption and 30 per cent improvement in energy efficiency (compared to projections).[21] The EC proposed revision of the emissions trading scheme, diversification of supply, new national plans and greater interconnection. With regard to its latest commitments, the EU has proposed to reduce 80–95 per cent of greenhouse gas emissions by 2050, compared to 1990 levels. An 'Energy Roadmap 2050' is currently in place for the assessment of possible pathways and challenges for the EU to achieve targets, guaranteeing security of supply and competitiveness.

All the above-mentioned plans are in line with the EU's energy law trilemma – security of supply, competition and sustainable development. The dissemination of renewable energy is crucial to upholding these policy goals and achieving its ambitious targets of reducing greenhouse gas emissions. Because of the opportunities that RE technologies offer in terms of market development and energy efficiency, the EU has seen its member states adopt multiple regulatory and voluntary support mechanisms for its deployment. This leads to the next section on the specific regulatory support mechanisms implemented in the EU thus far.

Support mechanisms for deployment of RE in the EU

Currently, each member state has its own set of support mechanisms (Figure 16.3). In general, countries have a main support mechanism (FIT or quota obligation) (Table 16.2) combined with de-risking financial policies (Figure 16.3). These sets of mechanisms have changed a lot over time and are different depending on the sector (Figure 16.4).

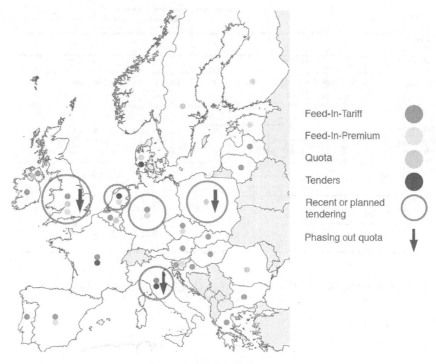

Figure 16.3 EU member states' regulatory support schemes. Data source: 22

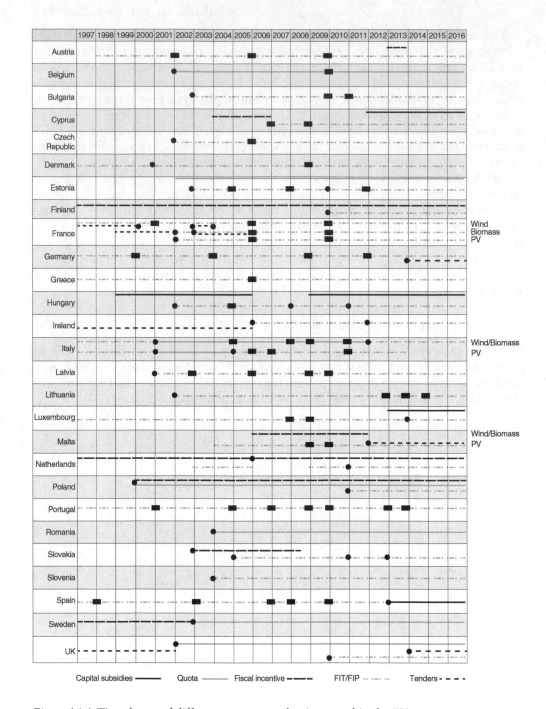

Figure 16.4 Time frame of different support mechanisms used in the EU.

Table 16.2 RES support mechanisms in EU, including financial and fiscal incentives

	RES-E support schemes, EU				
	Regulatory policies			Incentives	
	FIT/ FIP	Quota obligation	Tenders	Capital subsidy	Fiscal incentives
Austria	x				x
Belgium		x			
Bulgaria	x				
Cyprus	x	x			
Czech Republic	x			x	x
Denmark	x				
Estonia	x				
Finland	x				x
France	x		x		
Germany	x		x		
Greece	x				
Hungary	x			x	
Ireland	x		x		
Italy	x	x			
Latvia	x				
Lithuania	x				
Luxembourg	x			x	
Malta	x		x		x
Netherlands	x				x
Poland	x	x			x
Portugal	x				
Romania		x			
Slovakia	x				x
Slovenia	x				
Spain	x			x	
Sweden		x			x
UK	x	x	x		

According to Figure 16.4, a combination between price-based mechanisms and quantity-based mechanisms can be seen, as in the UK. In addition a trend can be noticed with respect to shifts between FIT to FIP and from FIT to Tenders, as in Italy. Finally, it can be seen that there is clearly no possible one-size-fits-all mechanism or set of mechanisms, and different instruments are combined depending on different technologies, under different states of maturity and facing distinct local rules, regulatory states and policy goals.

Criteria and indicators for performance

There are two types of efficiency criteria to assess the success of RES support schemes: dynamic and static.

STATIC EFFICIENCY

The simplest ways to measure the static efficiency or cost-effectiveness are: calculate the cost of installed capacity (USD/MW) as the report between cost of production and number of MW produced, or calculate the cost of electricity generation (USD/MWh) as a report between cost of production and total electricity generated in MWh. The results will vary significantly depending on whether the total costs, or only the technology costs, are considered.

The International Energy Agency (IEA) created the remuneration adequacy indicators (RAI) that help to compare remuneration levels between countries. The factors required are: payment schedule of incentives, influence of resource endowment and interaction between incentive levels and system prices.

The payment schedule of incentives expresses the remuneration as a levelised return over a fixed period, discounted at a fixed rate. The interaction between incentive levels and system prices determines whether the prices are influenced by other factors.

Two types of remuneration are considered: *up-front* (e.g. cash rebates and tax incentives) and *per MWh* (wholesale market revenues, certificate revenues, FIT, FIP). These remunerations are levelised to 20 years, considering a discount rate of 6.5 per cent using the following equation:

$$NPV = \sum_{t=1}^{n} \frac{Remuneration_t}{(1 + i)^t}$$

$$A = \frac{i}{1 - (1 + i)^{-n}} \times NPV$$

Where:

NPV = Net present value
i = Interest rate
t = Year
n = Payback time
A = Annualised remuneration level.

Then the remuneration is converted into USD/MW (capacity) and multiplied by the full-load hours of the considered country.

Generally, the literature only takes into account the technology costs. The IEA recommends considering the total costs in the calculation in order to estimate the costs for the consumers and to avoid windfall profits for producers.[23] The total costs indicator is defined by the IEA as 'the amount of additional annual premiums that are paid for the additional generation produced in a given year'.[24]

DYNAMIC EFFICIENCY

The dynamic efficiency approach focuses on the impact on technological diversity, investments in R&D, learning effects and technological competition. It takes into account how a policy can have an impact in the medium to long term on the improvements and cost reductions of the technology targeted. To measure this, Intelligent Energy Europe (IEE) considers two main indicators: one is the degree of deployment of more expensive or relatively immature technologies, measured as percentage deployment of different technologies with respect to potentials by country, and the other is the development of investment costs over time (€/kW). Another method is to use the learning curves to measure the potential reduction in costs a policy can induce. The method consists in determining to what extent a technology can see its costs reduced, what investments need to be made to do so, and if it is profitable in the long run.

Analysis of support mechanism for wind

Over the past decade, wind energy has developed tremendously and in 2015 reached a cumulative installed capacity of 143 GW in the EU, with 131 GW onshore and 11 GW offshore. Figure 16.5 shows the wind power installed by each member state at the end of 2015.

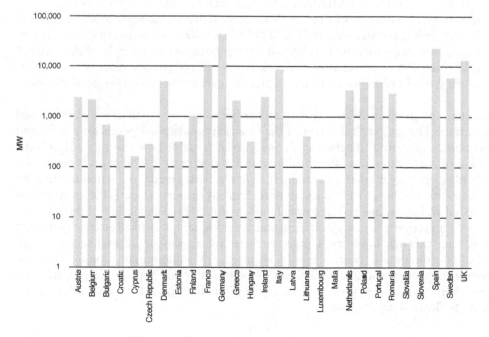

Figure 16.5 Wind installed capacity of EU member states. Data source: 25

Two case studies are discussed for this region: the UK and Denmark. They were chosen due to the high wind capacity installed.

In order to determine the maturity of offshore wind in these case studies, the deployment status indicator is considered. Table 16.3 shows a number of characteristics for wind energy in the UK and Denmark.

A number of European countries have high potential for offshore wind (France, the Netherlands, Sweden, Spain, Finland, Ireland, Italy). In order to develop their markets, these countries could benefit from the experience of the UK and Denmark. Unfortunately, some countries with high targets are not currently implementing sufficient support to reach these targets. For example, France had no installed capacity in 2012 when it had planned 667 MW. This is mainly due to administrative barriers, with consenting procedures taking on average six years in France, when it only takes two years in Germany. To overcome this problem, since October 2013 France has begun to test the one-stop-shop approach of the UK and Denmark. In other countries such as Italy, the main problem is the stability and level of the support mechanisms. A higher and more stable support system, like those implemented in the UK and Denmark, would be preferable.

Process to liberalise and integrate the European electricity markets

Table 16.4 summarises the historical time frame and geographic extent of European electricity markets. It was concluded the most important were:

* *the two major directives* (Directive 96/92/EC and Directive 2003/54/EC);
* *the comprehensive Third Legislative Package* (Directive 2009/72/EC, Regulation 713/2009/EC and Regulation 714/2009/EC), which was enacted to dismantle vertically integrated monopolies, liberalise the production and supply of electricity,[26] guarantee non-discriminatory network access to third parties, place wider emphasis on cross-border interconnections and mitigate barriers to cross-border trade.[27]

The extent of European electricity markets has been discussed extensively over the past few years.[28] The geographic extent of the electricity markets depends on a sufficient cross-border capacity and network of connections. A major goal that the EC has is the creation of a common and integrated market for electricity.[29] The main benefits of a single EU electricity market are: more efficient electricity trade, reduction in the need for electricity reserves, availability of more demand resources, increased competition and social welfare gains, reduced imbalances and increased flexibility of the systems, increased investment in energy-efficient technologies, and improved electricity market functioning through liquidity. However, there are a number of challenges: a high level of coordination between agents is required, transmission constraints can cause network congestion and uncoupling of European wholesale markets, large funding is required, and congestion problems can occur if the infrastructure required for traded electricity flows is insufficient.

Table 16.3 Case studies: the UK and Denmark

	UK	Denmark
Installed capacity growth (in MW) (The Wind Power 2014)[a]		

	UK		Denmark	
Total wind as share of electricity consumption in 2012 (Eurostat 2012)[b]	6%		27%	
	Onshore 4%	Offshore 2%	Onshore 21%	Offshore 6%

	UK	Denmark
Targets by 2020 (Beurskens and Hekkenberg 2011)[c]	12,990 MW	1,339 MW
Support mechanisms	Quota obligation with green certificates + FIT for projects < 5MW	FIT
Average administrative lead time (European Wind Energy Association 2010)[d]	26.87 months	31.81 months
Average grid access lead time (European Wind Energy Association 2010)[e]	8.36 months	2.01 months
Grid connection costs	Project developer	TSO
Market preparedness indicator (IRENA 2014)[f]	68	83
Equity indicator (Belgian Offshore Platform 2013)[g]	NSS = 74.2€/MWh GPI = 55.8€/MWh	NSS = 57.1€/MWh GPI = 64€/MWh

Notes

a The Wind Power (2014). *Denmark*. Accessed January 2015 from The Wind Power: www.thewindpower.net/country_en_6_denmark.php

b Eurostat (2012). *Statistics in Focus 44/2012*.

c L. W. M. Beurskens and M. Hekkenberg (2011). *Renewable Energy Projections as Published in the National Renewable Energy Action Plans of the European Member States*. Covering all 27 EU Member States, European Environment Agency. Accessed January 2014 from www.ecn.nl/docs/library/report/2010/e10069.pdf

d European Wind Energy Association (2010). *Wind Barriers – Administrative and Grid Access Barriers to Wind Power*. Accessed April 2016 from www.ewea.org/fileadmin/files/library/publications/reports/WindBarriers_report.pdf

e European Wind Energy Association (2010). *Wind Barriers – Administrative and Grid Access Barriers to Wind Power*. Accessed April 2016 from www.ewea.org/fileadmin/files/library/publications/reports/WindBarriers_report.pdf

f IRENA (2014). *Evaluating Renewable Energy Policy: A Review of Criteria and Indicators for Assessment*. Accessed June 2016 from www.irena.org/menu/index.aspx?mnu=Subcat&PriMenuID=36&CatID=141&SubcatID=379

g Belgian Offshore Platform (2013). *Benchmarking Study on Offshore Wind Incentives*. Accessed May 2016 from www.belgianoffshoreplatform.be/en/publications

Table 16.4 Historical time frame for European energy market integration and liberalisation (main focus: electricity)

Time frame	Description
By 1921	First network extension for electricity transportation was initiated.
1921–1922	Initiated project to liberalise and integrate European energy market.[a]
1925	Foundation of a first supervisory body for transmission lines.
1939–1949	The electricity sector in most European countries became owned or at least largely controlled by their respective governments.
1957	The Treaty of Rome planned liberalisation for all commodity markets including the energy market.[b]
1959	The exchange of electricity became liberalised, allowing national energy companies to engage more flexibly in corresponding transactions.[a]
1987	Single European Act – a step to abolish state-owned national monopolies.
1990	Liberalisation of wholesale electricity markets in the EU.
1992	Maastricht Treaty strengthened creation of free cross-border trade in electricity.
1996	First Electricity Directive 1996/92/EC – liberalisation of electricity market.
2000	Florence Forum – reached early agreements concerning market-based mechanisms for congestion management.[c]
2003	Second Electricity Directive 2003/54/EC – common rules of the internal electricity market.
2004	EU strategy on electricity market integration.[c]
2005	Adopted second internal market package for electricity.
2006	Electricity Regional Initiatives were launched.
2009	Third Electricity Package.
2011	EU adopted new stringent rules on wholesale electricity trading and sets target of 2014 for the completion of the internal electricity market.
2014	Multiregional price coupling in North Western Europe.
2015	New EU Regulation 2015/1222 establishing a guideline on capacity allocation and congestion management.

Notes
a G. Zachmann (2013). *Electricity without Borders: A Plan to Make the Internal Market Work*. Brussels: Bruegel Blueprint. Accessed April 2016 from http://bruegel.org
b J. Percebois (2008). Electricity liberalisation in the European Union: Balancing benefits and risks. *The Energy Journal*, 29 (1), 1–19.
c G. Squicciarini, G. Cervigini, D. Perekhodtsev and C. Poletti (2010). *The Integration of the European Electricity Markets at a Turning Point: From the Regional Model to the Third Legislative Package*. Series/Report No: EUI RSCAS; 2010/56; Florence School of Regulation, Italy. Accessed January 2014 from http://hdl.handle.net/1814/14294

South America

Overview

Like the EU, South America (SA) is composed of countries with very distinct characteristics; though, unlike the EU, countries in South America do not enjoy the same functional integration that fuses the economic, political, social and environmental aims of EU member states, through supranational governance and a pooling of sovereignty,[30] which is focused on promoting greater interdependencies and policy coordination among member states. Hence, while in the EU there are institutions governed by principles and procedures enshrined in EU law, with the capacity to set the general policies and guidelines for renewable energy at regional level to be applied at state level, in South America, regardless of regional integration efforts (e.g. MERCOSUR, UNASUR), no directly elected Parliament has yet enacted regional laws on energy issues. As such, all renewable energy initiatives are restricted to national rules and policy goals in South America, while in the EU they follow the general policies and guidelines for renewable energy at the EU level. Consequently, there are also no transnational agencies overlooking energy issues in South America, as in the EU, where there is the Agency for the Cooperation of Energy Regulators (ACER). ACER's missions include the creation of a more competitive, integrated market, focused on guaranteeing free movement of energy across borders, and transporting new energy sources to benefit consumers' choice and provide more efficient energy infrastructure.[31]

In South America regional integration took off with the creation of MERCOSUR – the Common Market of the South – through the Treaty of Asuncion in March 1991 by a joint decision of Argentina, Brazil, Uruguay and Paraguay. It is a customs union and a free trade area.[32] As the leader in regional integration initiatives, Brazil views MERCOSUR as part of a larger South American integration vision. This pushed negotiations on the South American Free Trade Area and culminated in the creation of the South American Community of Nations in 2004, renamed UNASUR in 2007. It has 12 member states with the collective objective of addressing regional cultural, social, economic and political issues, including areas such as security, infrastructure and energy.[33] Under the 19th Meeting of the Energy Expert Group of the South American Energy Council held on 21 August 2015, the general secretary of UNASUR emphasised the region must change its methods of use of renewable energy generation, the work of the Committee being to standardise energy policies that will allow for this to become a real strength for all nations.[34] While there are no agreed regional policies aligning action across countries in South America, each jurisdiction follows its own strategies and pursues its individual policy goals in terms of renewable energy deployment. As with the EU, the administrative and regulatory frameworks of each country are distinctly tailored to their electricity systems, making it hard to create greater convergence between each country's policy systems.

Energy is acknowledged as one of the most strategic factors of the region.[35] Most of the regulatory reforms happening in the electricity industry in South America are the result of having to attract enough investment to attend to increasing demands and growth.[36] As a place with very favourable conditions for the dissemination of non-conventional renewable sources because of its rich hydropower potential, significant solar resources, availability of biomass, very large reserves of natural gas power resources and persistent wind flows, the region has experienced a rise in the participation of these sources of energy (Figure 16.6).[37]

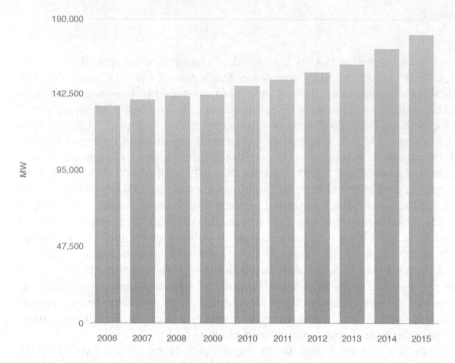

Figure 16.6 Total renewable energy in South America. ^{Data source: 38}

Differences lie in terms of prevailing generation technologies, availability of fossil fuels, and hydropower potential.[39] The total electricity production for each country discriminated by source in Figure 16.7 shows that their sources vary, but there is a predominance of hydropower in almost all markets, consequently making it quite suitable for combining with intermittent non-conventional renewable sources. As such, RES currently serves as a valuable alternative for these countries to diversify their energy mix, attend to their continuous increasing demands, universalise access and comply with their newly assumed compromises under the international climate change framework. Since the Paris Agreements, all countries should now contribute to the reduction of greenhouse gas emissions. Thus, regulatory models that promote a greater dissemination of these sources are key to allow these countries to respond to the continuous increase in electricity demand and help keep global temperatures below the two degrees agreed on in Paris in 2015.

Power sector reforms

Countries of the region have gone through multiple power sector reforms. In 1982 Chile issued its Electricity Act, which aimed for the liberalisation of its market. It introduced competition and required the unbundling of vertical-integrated utilities.[40] Its pioneering reform became a model for several South American countries such as Argentina (1992), Peru (1992), Colombia (1993) and Brazil (1994), which restructured their power sectors following, to a greater or lesser extent, the guidelines of Chilean liberalisation.[41] The conceptual background of these reforms lay in

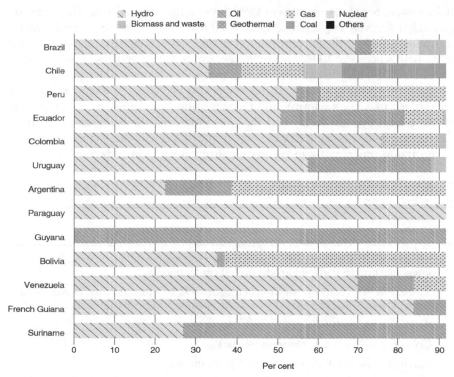

Figure 16.7 Electricity production of countries in South America in 2014. ^{Data source: 42}

the marginal pricing theory, where the optimal generation mixes were to be driven exclusively by economic signals resulting from the relationship between supply and demand and the cost of supplying one extra unit.

Several regulators complemented such policy from the very beginning with some sort of capacity mechanism (capacity payments, capacity markets, or both), in order to provide investors with a further incentive.[43] Chile adopted such a structure from the very beginning in 1982. Other countries to follow it were Colombia, Argentina and Peru. The main exception to this in the region was Brazil, because it relied on a quantity scheme; according to its rules, power purchase agreements were to be used to back up the necessary load to attend to what was once 85 per cent of forecasted demand, and now is 100 per cent.

For a number of reasons (ill designed and regulatory interference), the capacity mechanisms did not work as expected and the expansion of the electricity supply was not able to keep up with the fast-paced growth of demand.[44] In Chile, for example, trouble began in 1999 as a consequence of its worst drought, which put to the test its whole regulatory model. Meanwhile, in Brazil, investments did not increase as expected with the market reform introducing marginal pricing in 1996. Consequently, it had to use more of its hydro capacity to support the lack of investment between 1998 and 2000. When the severe drought happened in 2001 it had no means of securing supply, and rationing became the only alternative. In the face of issues that resulted not only in the shortages described above, but also in the narrowing of reserve margins, and higher prices, regulators were forced to implement new reforms.

Initial discussions began in Colombia in 1999 through the proposal 'reliability options', which was only effectively implemented later in 2006.[45] The backbone of most reforms in South America involved the implementation of all sorts of long-term auctions. Brazil launched its new scheme in 2004. Similar long-term auction mechanisms were implemented in Chile (2005) and Peru (2006). Hence, the main reforms in the region involved the transition from vertically integrated monopolies to liberalised markets and competition (encouraged by Chile's reform); and the transition from capacity mechanisms to long-term electricity auctions, inspired by the Colombian and the Brazilian experiences. Table 16.5 has a detailed description of the time frame of sector reforms in the region and the different support programmes and schemes existing for RES in each country.

Table 16.5 Historical time frame of energy market reforms and support schemes for RES in South America

Time frame	Description
1982	Chile issued its Electricity Act which considered the unbundling of vertical integrated utilities and the introduction of competition.[a]
1991	MERCOSUR – Treaty of Asuncion in March 1991 by a joint decision of Argentina, Brazil, Uruguay and Paraguay. It is a customs union and a free trade area.[a]
1992	Argentina and Peru followed Chile's reform.[a]
1993	Colombia follows with reform of its market.
1994	Brazil introduces reform inspired by Chile's model.
1999	New mechanism based on 'reliability options' was proposed in Colombia.
2001	Colombia enacts Law 697 to promote the efficient and rational use of energy and alternative energies.[b]
2002	PROINFA Programme launched in Brazil to support RE through price-based FIT.[b]
2004	Brazil introduces new scheme based on long-term auction mechanism with central role of guaranteeing security of supply.
	South American Community of Nations – UNASUR composed of 12 member states.
2005	Chile implements long-term auction mechanism.[b]
2006	Peru launches its long-term auctions.
	Colombia launches a scheme based on the reliability-option proposal.
	Argentina launches the Energy Plus programme which resembles long-term electricity auctions (no auction has yet been carried out under this scheme).
	Uruguay's fiscal incentives for renewables under Auctioning Decree 77–2006.
2008	Brazil implements its first Reserve Energy Auction oriented exclusively towards non-conventional RE.
	Peru launches support scheme for RE and considers premium on top of the market price to guarantee a 12% rate of return (competitive bidding).
	Chile introduces its main support scheme for RE – Quota.
2009	Chile Regulatory Framework for Solar Water Thermal (Law 20,365).[b]
2010	Colombia adopts Resolution 180919 with programme for rational use of energy, defining specific targets and penetration of RE: 3.5% by 2015 and 6.5% by 2020.

Time frame	Description
	Argentina implements Renewable Energy Generation Program (GENREN) requiring its state-owned utility ENARSA to contract RE (1,015 MW).
	Uruguay implements Feed-In-Tariff for promotion of renewable energies according to Decree 354.[b]
2011	Colombia allows wind projects to take part in Reliability Charge Auctions.
	Peru enacts new regulations of electricity generation from RE.
	Uruguay conducts auctions of up to 150 MW.
2012	Brazil net metering for Distributed Generation.[b]
	Chile's support for non-conventional renewable energy development programme (RD&D funding).
2013	Ecuador enacts CONELEC 001/13, implementing Feed-In-Tariff.[b]
	Uruguay implements auctions and FIT for private generation of photovoltaic energy.[b]
2014	Brazil's Wind Turbine Component Tax Exemption.[b]
	Chile's net metering regulation on distributed generation.[b]
	Bolivia amends its Electricity Law to establish the Feed-In-Tariff.[b]
2015	Bolivia declares agenda with objectives linked to renewable energy development.

Notes
a P. Mastropietro, C. Batlle, L. A. Barroso and P. Rodilla (2014). Electricity auctions in South America: Towards convergence of system adequacy and RES-E support. *Renewable and Sustainable Energy Reviews*, 40, 375–385.
b International Energy Agency (IEA) (2016). *Policies and Measures Databases*. Accessed January 2014 from www.iea.org/policiesandmeasures

Support mechanisms for deployment of RES in South America

RES energy support mechanisms have been present in the South American region for the past ten years. Currently, each country has its own set of support mechanisms (Figure 16.8). Generally, countries have a combination of several instruments, with a main support mechanism (FIT or Tenders) combined with financial de-risking policies (Figure 16.8).

The support schemes in Table 16.6 have greatly changed over time in the region as shown in Figure 16.9. The subsections below will look closer into the two largest markets for renewable wind energy in the region: Brazil and Chile (see also Table 16.7).

We have chosen these two countries because they are the two with the highest installed capacity for wind in South America, both having very different schemes in place to support the deployment of wind. Whereas in Brazil it is a centralised system, with the government making all decisions with regards to tendering, in Chile it is decentralised and the distribution companies are responsible for launching the auctions. In Chile, the main support scheme is quotas, where the distribution companies have to cover a progressive percentage of their load through renewable technology (up to 10 per cent in 2024). Companies who exceed this are entitled to trade their surplus in the market. In Brazil there is no quota mechanism in place and the distribution companies are subject to the government's decision when it comes to the deploying of renewable energy to attend to its demand.

Feed-In-Tariff

Quota

Tenders

Figure 16.8 Support schemes in South America.

Table 16.6 Support schemes in South America

	RES-E support schemes, South America						
	Regulatory policies				*Fiscal incentives*		
	FIT	*FIP*	*Quota obligation*	*Tenders*	*Capital subsidy*	*Reduction in taxes*	*Tax credits*
Brazil				x	x	x	x
Chile			x	x	x	x	x
Peru	x			x	x	x	x
Ecuador	x			x	x	x	
Colombia	x		x			x	x
Uruguay	x			x	x	x	x
Argentina	x			x	x	x	x
Paraguay						x	
Guyana							x
Bolivia	x				x	x	

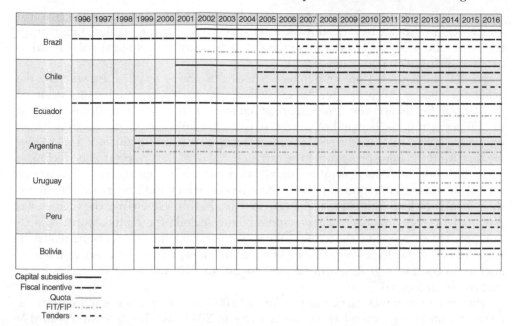

	1996	1997	1998	1999	2000	2001	2002	2003	2004	2005	2006	2007	2008	2009	2010	2011	2012	2013	2014	2015	2016
Brazil																					
Chile																					
Ecuador																					
Argentina																					
Uruguay																					
Peru																					
Bolivia																					

Capital subsidies ———
Fiscal incentive — — —
Quota ———
FIT/FIP ·· ·—· ··
Tenders · — — ·

Figure 16.9 Time frame of different RES support mechanisms implemented in South America.

BRAZIL

The 'Proinfa' programme, launched in 2002, was the first scheme adopted in Brazil that created incentives for the development of RES. It was essentially a FIT designed to contract until 2006 some 3,300 MW of wind, biomass and small hydro. Eletrobrás, a holding power utility owned by the Federal Government, is responsible for buying the energy produced by all plants that are part of the programme under 20-year power purchase agreements. Due to the poor performance of the projects and lack of economic signals for greater efficiency and technological improvements, Proinfa has received a lot of criticism,[46] owing to the numerous implementation issues and security of supply issues resulting from most projects being delayed.

The auction-based approach implemented in Brazil through its power market reform in 2004 is currently the main tool used in the country to foster RES. The reform was conceived to address concerns regarding the establishment of a stable regulatory framework to attract investments in the expansion of the generation system, ensure reliable supply levels and provide affordable tariffs resulting from competitive bidding. Aimed at achieving reasonable tariff rates, its new power market reforms require that: (a) all energy purchases by distributors must be made through auctions, based on the lowest rate; (b) contracting is realised through the pooling system of a regulated market; and (c) load contracting is separated into two types of transactions, by quantity of energy or by availability. Under regulated power purchase agreements by quantity of energy, the generator undertakes to provide a certain volume of physical energy and, thus, assumes the risk of supply of energy being affected by hydrological conditions or low water levels of reservoirs; in which event the producer is required to purchase electricity in the short-term market to meet the volume agreed.

Under regulated power purchase agreements by availability of energy, the generator undertakes to provide buyers with a specific capacity under the regulated market. In the latter, the producer's revenue is guaranteed and any hydrological risks are allocated to distribution companies.

For RE the regulated power purchase agreements are by availability, whereby the generator undertakes to provide buyers with a specific capacity under the regulated market. All distribution companies are required to estimate their energy needs to attend to 100 per cent of the demand of its captive customers over a five-year horizon. These estimates serve to indicate the need for constructing new plants in advance, thus allowing time for the bidding procedure to be carried out and for the projects to be developed. Since 2005, all electricity generation, distribution and trading agents, as well as free or special customers, are required to notify Ministry of Mines and Energy (MME) about their market or load estimates, as the case may be. In this framework, the regulator is essentially a system planner. Their responsibilities include the establishment of auction procedures, including the terms of the power purchase agreements being tendered. Although all technologies compete in Brazilian auctions, the government can call an auction to contract a government-selected volume of RES and select the technologies that can be negotiated.

The first auction was carried out in August 2008 to contract new energy from the cogeneration of sugarcane bagasse for delivery in 2011 and 2012. Some 2,400MW (gross capacity) was acquired in 15-year contracts for an average price of 80USD/MWh. The first wind auction was held in 2009. A 20-year energy contract with delivery starting in 2012 was offered to investors, with a very specific accounting mechanism designed to provide investors with a fixed payment and incentivising/penalising production above/below a given energy threshold.[47]

Brazil has 12 years of experience in auction mechanisms. Since its market reform in 2004 it has increased its installed capacity.

CHILE

Similar long-term auctions were implemented in Chile in 2005. Different from the Brazilian case, it is the distributors who organise and run their auctions, instead of the regulator. Distribution companies hold long-term energy contract auctions to supply their regulated consumers in which no technology discrimination is applied. In 2009 a wind farm won a 275GWh/year 15-year energy contract for a price of 93USD/MWh.[48]

Following a very different path than its neighbours, electricity regulation was modified in 2008 with the introduction of a quota system, which determines that at least 10 per cent of the energy traded by generators should be produced by RES.[49] It starts with an obligation of 5 per cent for the period between January 2010 and 2014, followed by an increase of 0.4 per cent annually until 10 per cent is reached by 2024.[50] Non-compliance leads to a fine of 28USD/MWh that, if repeated in a three-year period, can sum up to 42USD/MWh.[51]

In order to comply with these objectives, a strong volume of RES investment will have to take place over the next few years. Mini hydro (hydro plants smaller than 20MW), wind and biomass are the most economically attractive alternatives for the country. The main support scheme is the quota and, very differently from Brazil, it does not rely on long-term auctions to support the penetration of non-conventional renewable energy into the power sector.

Table 16.7 Case studies: Brazil and Chile

	Brazil	Chile
Installed capacity growth wind power (in MW)		
Installed capacity in 2015	8,715	904
Total TWh wind as share of electricity production	12.2TWh (MME 2016)	1.87TWh (CDECSIC 2016)
Targets	6GW by 2019	20% non-hydro by 2025
Support mechanism	**Tenders** A1, A3, A5 and Reserve auctions, where reserve auctions have been largely used with a specific product designed for wind power generation	**Quotas/Auctions** Distribution companies need to guarantee that a certain percentage of its purchases comes from non-conventional renewable generators for each calendar year
Administrative lead time	According to each individual permit, observed the terms applicable to tenders A1 (1 year), A3 (3 years) and A5 (5 years) and reserve auctions (usually 3 years)	Phase 1 = installation application to distribution company that has to be responded to in 15 days, so that the interested party can develop and design connection engineering and submit Network Connection Application Phase 2 = distribution company has two months to respond to Application, which, if positive, has an 18-month validity period
Grid access lead time	Coincides with above date in which it has to start operations	18 months upon acceptance by the distribution company of the Network Connection Request
Grid connection costs	Developer bears costs. Benefiting from at least a 50% discount	Developer bears costs. Charge exemption is calculated as a proportional adjustment ranging from 100% for RE projects of 9 MW and 0% for 20 MW or more

Conclusions

The main conclusions are:

- The European Union has done significant work for the deployment of RES. However, there is no single best support mechanism that will work in all member states, because of the different resources and internal markets. Indicators can help policymakers to find the best solution for a country.
- Offshore wind shows some great results in some EU countries. Support mechanisms are evolving and as a result renewable energy will be more competitive, and this will result in phasing out the support instruments once RES is introduced to the real market.
- An important issue for European cooperation is the grid extension. There is an obvious but expensive need to reinforce the grid. In order to facilitate the creation of an internal European energy market and to ensure the maximisation of benefits, the European Commission Directive 2009/28/EC has created cooperation mechanisms that enable the member states with less potential to benefit from the energy produced in other countries to reach their targets. The price of this transfer depends on a number of parameters, from the energy markets of each member state to the time left until the targets must be reached. It also depends on the production of the other European countries.
- Since liberalisation started in the South American region, countries have opted for a combination of regulatory support schemes to attract investments in generation. Whereas in the 1980s the majority of countries opted for market-oriented schemes based on short-term marginal pricing, nowadays long-term auctioning is the most used mechanism in the region, followed by FIT. Chile is an exception to the region, with a combination that has included quota schemes since 2010. Finally, although auctioning is the most common scheme in South America, the many different design elements of these mechanisms make them particular to each country and have a significant impact on the auction results. Long-term electricity auctions have shared common objectives: fostering the entrance of new capacity by fixing in advance part of the investor's remuneration, thus hedging their risk.
- Today, renewable energies are starting to be competitive and this will result in a phasing out of the support instruments in order to introduce progressive renewable energy resources to the market. To do so, some important improvements will be required in order to allow efficient cooperation between countries within regions.

16.4 Future perspectives on cooperation mechanisms

One of the many cooperation mechanisms is that of the statistical transfer. This involves the statistical transfer of the host country, with a considerable advantage in terms of possessing natural resources, to the off-taking country, to help that country achieve their targets. Another mechanism pertains to the shared projects between member states. There may be certain considerations, when it comes to this particular cooperation mechanism, that may cause issues in the functionality of the mechanism as a whole. For instance, how to choose the member state that will be responsible for the payment, monitoring and support of the project. Overcoming these potential problems will be crucial in the development and sustenance of an efficient renewable energy market.

Notes

1 D. M. Newbery, L. Olmos, S. Rüster, S. J. Liong and J. M. Glachant (2011). Public Support for the Financing of RD&D Activities in New Clean Energy Technologies. *A Report for the European Commission, Directorate General Energy within the EC FP*, 7.

2 C. Batlle and L. A. Barroso (2011). *Review of Support Schemes for Renewable Energy Sources in South America*. Department of Economics, MIT, and Sloan School of Management. Accessed June 2016 from http://ceepr.mit.edu/files/papers/2011-001.pdf

3 Ecofys, Fraunhofer, TU Vienna EEG and Ernst & Young (2011). *Financing Renewable Energy in the European Energy Market*. Accessed June 2016 from https://ec.europa.eu/energy/sites/ener/files/documents/2011_financing_renewable.pdf

4 Batlle, C. and Barroso, L. A. (2011). *Review of Support Schemes for Renewable Energy Sources in South America*, A Joint Center of the Department of Economics, MIT Energy Initiative, and Sloan School of Management Accessed June 2016 from http://ceepr.mit.edu/files/papers/2011-001.pdf

5 R. Haas, W. Eichhammer, C. Huber, R. Langniss, P. Menanteau, P. E. Morthorst, A. Martins, A. Oniszk, J. Schleich, A. Smith, Z. Vass and A. Verbrüggen (2004). How to promote renewable energy systems successfully and effectively. *Energy Policy*, 32, 833–839.

6 P. Mastropietro, C. Batlle, L. A. Barroso and P. Rodilla (2014). Electricity auctions in South America: Towards convergence of system adequacy and RES-E support. *Renewable and Sustainable Energy Reviews*, 40, 375–385.

7 Ecofys, Fraunhofer, TU Vienna EEG, Ernst & Young (2011). *Financing Renewable Energy in the European Energy Market*. Accessed April 2014 from https://ec.europa.eu/energy/sites/ener/files/documents/2011_financing_renewable.pdf

8 M. Ragwitz (2010). Developments and achievements of Feed-In Systems – Key findings from an evaluation conducted for the IFIC. In *8th Workshop of the International Feed-In Cooperation (IFIC), Berlin.*

9 Batlle, C. and Barroso, L. A. (2011). *Review of Support Schemes for Renewable Energy Sources in South America*, A Joint Center of the Department of Economics, MIT Energy Initiative, and Sloan School of Management Accessed June 2016 from http://ceepr.mit.edu/files/papers/2011-001.pdf

10 Ibid.

11 Ibid.

12 A. Verbruggen and L. Volkmar (2012). Assessing the performance of renewable electricity support instruments. *Energy Policy*, 45, 635–664. Accessed February 2014 from www.sciencedirect.com/science/article/pii/S0301421512002194

13 P. Mastropietro, C. Batlle, L. A. Barroso and P. Rodilla (2014). Electricity auctions in South America: Towards convergence of system adequacy and RES-E support. *Renewable and Sustainable Energy Reviews*, 40, 375–385.

14 Batlle, C. and Barroso, L. A. (2011). *Review of Support Schemes for Renewable Energy Sources in South America*, A Joint Center of the Department of Economics, MIT Energy Initiative, and Sloan School of Management Accessed June 2016 from http://ceepr.mit.edu/files/papers/2011-001.pdf

15 European Commission (2013). *Energy Challenges and Policy*. Accessed April 2016 from http://ec.europa.eu/europe2020/pdf/energy2_en.pdf

16 European Commission (2007). *An Energy Policy for Europe*. Accessed March 2016 from http://ec.europa.eu/energy/energy_policy/doc/01_energy_policy_for_europe_en.pdf

17 European Commission (2010). *Europe 2020, a Strategy for a Smart, Sustainable and Inclusive Growth*. Accessed March 2016 from http://eur-lex.europa.eu/procedure/EN/199842

18 Ibid.

19 European Commission (2015). *Annex to the Report of the Commission to the European Parliament, the Council, the European Economic and Social Committee and the Committee of the Regions – Renewable Energy Progress Report*. Accessed June 2016 from https://ec.europa.eu/transparency/regdoc/rep/1/2015/EN/1-2015-293-EN-F1-1.PDF

20 Eurostat (2015). *Energy Production and Imports*. Accessed June 2016 from http://ec.europa.eu/eurostat/statistics-explained/index.php/Energy_production_and_imports

21 European Parliament and Council (2012). *Energy Efficiency Directive 2012/27/EU*. Accessed April 2016 from http://eur-lex.europa.eu/legalcontent/EN/TXT/?uri=uriserv:OJ.L_.2012.315.01.0001.01.ENG

22 Ecofys, Fraunhofer, TU Vienna EEG and Ernst & Young (2011). *Financing Renewable Energy in the European Energy Market*. Accessed June 2016 from https://ec.europa.eu/energy/sites/ener/files/documents/2011_financing_renewable.pdf

23 Intelligent Energy Europe (2012). *RE-Shaping. D5 & D6 Report: Indicators Assessing the Performance of Renewable Energy Support Policies in 27 Member States*. Karlsruhe: IEE.
24 International Energy Agency (IEA) (2011). *Policies and Measures Databases*. Accessed from www.iea.org/policiesandmeasures
25 European Wind Energy Association (2013). *The European Offshore Wind Industry – Key Trends and Statistics 2012*. Accessed April 2016 from www.ewea.org/fileadmin/files/library/publications/statistics/European_offshore_statistics_2012.pdf
26 E. Pellini (2014). *Essays on European Electricity Market Integration*. PhD thesis, University of Surrey.
27 T. Jamasb and R. Nepal (2012). *Market Integration, Efficiency and Interconnectors: The Irish Single Electricity Market*. Edinburgh: Heriot-Watt University. Accessed June 2016 from http://papers.ssrn.com/sol3/papers.cfm?abstract_id=2025628
28 V. Böckers and U. Heimeshoff (2014). The extent of European power markets. *Energy Economics*, 46, 102–111.
29 T. Jamasb and R. Nepal (2012). *Market Integration, Efficiency and Interconnectors: The Irish Single Electricity Market*. [Online]. Edinburgh: Heriot-Watt University. Accessed June 2016 from http://papers.ssrn.com/sol3
30 R. Keohane and S. Hoffman (1991). Institutional change in Europe in the 1980s. In R. Keohane and S. Hoffman (eds), *The New European Community: Decision Making and Institutional Change*. Boulder, CO: Westview Press.
31 ACER (Agency for the Cooperation of Energy Regulators) (2016). *Regional Initiatives*. Accessed June 2016 from www.acer.europa.eu/en/The_agency/Pages/default.aspx
32 E. Lazarou (2013). Brazil and regional integration in South America: Lessons from the EU's crisis. *Contexto Internacional*, 35 (2), 353–385.
33 UNASUR (2015). *Union of South American Nations*. Accessed February 2016 from www.unasursg.org/en
34 Ibid.
35 Ibid.
36 P. Mastropietro, C. Batlle, L. A. Barroso and P. Rodilla (2014). Electricity auctions in South America: Towards convergence of system adequacy and RES-E support. *Renewable and Sustainable Energy Reviews*, 40, 375–385.
37 Ibid.
38 IRENA (2014). *Evaluating Renewable Energy Policy: A Review of Criteria and Indicators for Assessment*. Accessed June 2016 from www.irena.org/menu/index.aspx?mnu=Subcat&PriMen uID=36&CatID=141&SubcatID=379
39 P. Mastropietro, C. Batlle, L. A. Barroso and P. Rodilla (2014). Electricity auctions in South America: Towards convergence of system adequacy and RES-E support. *Renewable and Sustainable Energy Reviews*, 40, 375–385.
40 C. Batlle, L. A. Barroso and I. J. Pérez-Arriaga (2010). The changing role of the State in the expansion of electricity supply in Latin America. *Energy Policy*, 38 (11), 7152–7160.
41 Ibid.
42 Ibid.
43 Ibid.
44 P. Mastropietro, C. Batlle, L. A. Barroso and P. Rodilla (2014). Electricity auctions in South America: Towards convergence of system adequacy and RES-E support. *Renewable and Sustainable Energy Reviews*, 40, 375–385.
45 C. Vazquez, M. Rivier and I. J. Pérez-Arriaga (2002). A market approach to long-term security of supply. *IEEE Transactions on Power Systems*, 17 (2), 349–357.
46 P. Mastropietro, C. Batlle, L. A. Barroso and P. Rodilla (2014). Electricity auctions in South America: Towards convergence of system adequacy and RES-E support. *Renewable and Sustainable Energy Reviews*, 40, 375–385.
47 F. Porrua, B. Bezerra, L. A. Barroso, P. Lino, F. Ralston and M. Pereira (2010, July). Wind power insertion through energy auctions in Brazil. In *IEEE PES General Meeting* (pp. 1–8). IEEE.
48 C. Batlle, L. A. Barroso and I. J. Pérez-Arriaga (2010). The changing role of the State in the expansion of electricity supply in Latin America. *Energy Policy*, 38 (11), 7152–7160.
49 Ibid.
50 IEA (2016). *Policies and Measures Databases*. Accessed from www.iea.org/policiesandmeasures/
51 Ibid.

17 Energy resource use and geopolitics of trade

17.1 Introduction

The distribution of natural resource deposits across the global landscape is highly uneven. This can be demonstrated by the fact that only 15 of the world's states possess around 90 per cent of proven global oil reserves. Due to this uneven distribution, there is considerable difference between the patterns of utilisation of these resources in the various regions of the world.

The report by Verrastro *et al.*, *The Geopolitics of Energy Emerging Trends, Changing Landscapes, Uncertain Times*,[1] published in 2010 as part of the CSIS energy and national security programme, has been a great example of the premise that 'the relatively benign global energy situation that had persisted for the previous 15 years was masking emerging changes in both markets and international realignments'. One of the conclusions was that fossil fuels would continue to play a major role in the global energy needs for the next several years, with the Persian Gulf remaining the key marginal supplier of oil to the world, and with Europe's overreliance on Russian natural gas and US oil imports continuing to grow.

The authors provide a high-level overview of the relevant drivers that will dictate future trends in energy consumption, supply sources, geopolitical relationships, foreign policy and environmental choices.

In this chapter I will primarily focus on Europe and its reliance on Russian natural gas.

In line with the global trends, Europe experiences diversity in natural resources as well, from region to region and from country to country. Energy dependency varies highly among the Member States of the EU. A considerable percentage of the EU's energy imports are from politically unstable areas. Moreover, the EU also depends on Russia. The Russian imports include 35 per cent of oil, 30 per cent of coal, 35 per cent of uranium and 26 per cent of gas imported into the EU. In 2011 alone, Europe's dependency on energy imports amounted to 54 per cent. The region imports around 95 per cent of its uranium, 85 per cent of oil, 67 per cent of gases and 41 per cent of its solid fuels.

Even though Europe is rapidly moving towards increasing its exploitation of domestic energy sources in the future, it will continue to be highly dependent on energy imports in the coming years. According to the European Commission's projections, energy imports in the EU could increase to 55 per cent by 2030 and 57 per cent by 2050. The over-dependence on Russian energy imports puts the EU in a position of disadvantage in case any geopolitical tensions arise in Russia or any of the states through which the products are transmitted.[2]

Malta and Cyprus were the only two countries in the EU that did not rely on natural gas at all, according to 2013 statistics. Some of the countries where natural gas consumption was lower than 100 billion cubic feet per year included Estonia, Luxembourg, Bulgaria, Latvia, Sweden and Slovenia. Germany, Italy and the UK emerged as the biggest consumers of natural gas with yearly consumption surpassing 2,000 billion cubic feet. Other major consumers of natural gas included France, the Netherlands and Spain, which consumed more than 1,000 billion cubic feet per year each. The consumption in other states averaged between 100 and 600 billion cubic feet per year. Denmark and the Netherlands occupy a unique position in the region as they are the only two states that produce more locally than they consume. According to 2013 statistics, the Netherlands also imports natural gas.[3]

The complicated interlinkage between natural energy and other resources is highly influential on the use and distribution of natural energy in various regions of the world. This creates the need for effective regulation, policies and agreement among the producers and consumers. The interlinkages fall under what's commonly known as the nexus system. The system considers the interdependencies between food, water and energy as the primary resources for societies. It attempts to highlight the trade-offs and synergies that exist in an incorporated management of the resource systems.[4]

Water and energy share a complex and multifaceted relationship. On the one hand, water is essential for the extraction and processing of fossil fuel, growing biomass or biofuel, and the generation of electricity. On the other hand, energy is a necessary prerequisite for the transportation of water as well as treating and extraction of wastewater and drinking water respectively. None of these two sources should be considered separately. There is a need for a holistic approach that takes this interdependency into account. There is uncertainty around the ways to support nexus decision making due to the complexities of considering food, water and energy together as well as the sheer scope of each particular area. As a result, the regulations and policies frequently develop suboptimal signals to the concerns of national security, the economy and the environment. The policies created in a particular field focus on just two areas, and techniques to address the extensive interdependencies are few and far between. There is a need for systematic thinking to achieve this. However, this is something that cannot be easily translated into processes, particularly those pertaining to government policy.[5] According to Bazilian *et al.*, the positive aspects of an all-inclusive policy and a regulatory design would include better standards of public health, efficiency of resources, improved livelihood and economic stability. The negative aspects may include sub-par infrastructure design, commodity prices and environmental degradation.[6]

It is generally considered that national or regional energy policies are based on three pillars: economic competitiveness, energy security and environmental sustainability. This is especially true in the case of Europe, where every one of the aforementioned pillars is strengthened by a sole institution that includes separate directorates for the environment, competition, transport and energy. Any feasible energy policy is based on concessions aiming to find that crucial balance between potentially conflicting goals. Now, more than ever, there is a need to reduce or completely eliminate political conflicts when it comes to the energy nexus. A number of researches and simulations have been conducted to ensure the employment of the best policies in this area.

The Energy Research Centre of the Netherlands (ECN) and the Clingendael International Energy Programme designed a methodology and supply–demand index to analyse long and medium-term supply security. In supply, 30 per cent was meant for transportation and conversion, while 70 per cent was meant for basic energy supply.

Bazilian *et al.* analysed the impacts of an economic downturn on electrical supply security for the Slovenian power system over a period of four years – between 2004 and 2008. They developed an effective technique for quantifying risks and created a composite index for the task. There are three dimensions of this index: primary energy supply security, power system stability and environmental performance.

These dimensions are further broken down into sub-divisions and separate indices that are then aggregated into unique dimension indices. Data employed at dimensional level is already unique to the assessed electrical power system. Dimensional indices stand for contributions to the general composite index that are aggregated to one composite electrical supply security index. Nevertheless, this index does not allow benchmarking or direct comparisons among diverse power systems.[7]

Throughout history, human beings have always strived to find solutions to a wide range of complex and interrelated problems, which Lawford identifies as basic threats to human civilisation. Many of these issues are associated with food and water and the use, production and distribution of energy, particularly in developing states. Owing to the scope of the problem and the complexity of analysing these three resources at the same time, there is very little available research on the ways to encourage nexus decision making.

The nexus approach to food, water and energy depends on the views of the policy-makers. For instance, if they adopt the water view, then the inputs consist of water and energy. When we consider energy, water and bio-resources are the inputs, and food and resource requirements are, usually, the outputs. Irrespective of the case, the perspective influences policymaking. It depends upon the priorities of the respective ministry or institution or the breadth of analytical knowledge of the tools of the relevant experts and the support staff. Giurco *et al.* believe that very few people are experts in all three areas.[8]

17.2 Energy supply and political stability in Europe

Although the concept of energy security is frequently employed in academic literature and policy texts, the connotations are diverse due to the geopolitical and academic backgrounds of the users. The term energy security is usually used as a synonym for supply security, particularly by researchers studying energy security in the context of economics.

Supply security may be generally described as reliable, affordable and adequate energy supplies. In various other forums, the relationship between security and energy is perceived through the lens of energy as a contributor of security threats and conflicts. In analysis, these views are rarely integrated.

Kratochvil and Tichy identify three key events that have had a major role to play in increasing the significance of security supply. They include the Russian Ukrainian dispute and the subsequent decrease in gas supplies by about one-third in 2006, the complete interruption of gas supplies in 2009, as well as the electricity blackout of November 2006. The blackout, which affected almost half of EU Member States, was a result of the establishment of three interconnected lines with varying frequency.

The need for diverse transport routes contributed to the increased significance of the newer projects aimed at transporting gas to the EU from new sources and specifically from Northern Europe to Southern Europe. The 2010 regulations require all EU members to prepare preventive measures. It also built a foundation on rules for a joint and national preventive action plan design founded on emergency and risk assessment plans. According to Kratochvil and Tichy, the EU will also influence infrastructure investment planning by imposing a bidirectional flow.[9]

Although European countries have a diverse energy resource disposition and use, the continent has not experienced a major conflict among its states with regard to the use and exploitation of these resources. However, the inability of the available natural energy resources to satisfy the continent's needs have pushed Europe into developing an energy treaty that has resulted in more cooperation than conflict. The Energy Charter Treaty creates an outline for international cooperation among European nations and other developed nations with the objective of creating the energy potential of Eastern and Central European nations and guaranteeing energy supply security to the European Union. The energy efficiency protocol and associated aspects of the environment aim to enhance effectual energy policies that are compatible with appropriate development, encourage sound and more efficient use of energy, and enhance cooperation in the field of energy efficiency.[10]

17.3 The supply of oil and gas in Europe

Natural gas acts as the most important alternative energy source and arguably among the most geostrategically and geopolitically complex natural sources because, similar to oil, natural gas allocation is geographically uneven; it is hard to store since it exists as a gas; and the natural gas transportation infrastructure is strictly meant for natural gas transportation – gas pipelines – therefore contributing to geopolitical complexity. To alleviate the incurred risks as a result of natural gas characteristics, a number of measures that include development of long-term binding agreements, joint investment in pipelines, international consortiums and transit nations' compensation by gas price discounts or explicit transit fees have been employed so as to safeguard bargaining positions, avoid conflict and finally benefit from this complex geopolitical trade.

However, there have been serious conflicts in natural gas trade all through history. The representative and most classical case is the Russian natural gas trade via pipelines that pass via previous Soviet nations, particularly Belarus and Ukraine, to link Western Europe. Even after the collapse of the Soviet Union, Belarus and Ukraine benefited from Russian natural gas domestic-quasi prices, which were substantially lower as compared to Russian prices of exports to Western nations. However, as the demand for national gas went on increasing globally, and Gazprom – a natural gas enterprise owned by the state – chased more profitability, Russia started to demand higher prices from Belarus and Ukraine. This initiated continuous conflict and bargaining for over two decades, which was embodied in the announcing and planning of a number of alternative pipelines intended to bypass Belarus and Ukraine as transit nations.

Europe highly depends on Russia for its natural gas and oil supply among other national energy resources. This simply means that geopolitical instability along their supply line that includes Ukraine can highly affect natural gas supply in Europe and thus jeopardise the operation of the entire continent. In the past, Europe developed policies advocating for diverse means of natural energy transportation; however, with the current enactment of environmental law, some of those means may need to be reduced. In this regard, Europe will remain at high supply insecurity until it diversifies its sources further. This policy acts as one of the ways that can be applied to resolve future natural energy supply security threats. The second proposed policy involves the establishment of a European common gas purchasing vehicle – this simply advocates for energy union and monopolising energy purchase as a way of creating a strong bargaining power. However, this may take Europe back to the 1970s, before liberalisation of the energy market, which may be regarded as a backward step. The last proposed policy are sanctions focusing on the oil sector of Russia. This would weaken Russia's economic power and thus make it less controlling; however, the country can still use its product to control others in other, more severe ways: as a weapon.

Russia has for a long time been a threat to most nations that depend on it for gas and oil, especially former Soviet states that are regarded as being more dependent, poor and small. Russia has in the past employed its influence to punish its enemies and to reward its friends as a way of controlling the region. The country has shown that it can easily succeed in this by imposing high prices on those who dare defy it. In this regard, Russia has always stirred historical conflict especially between it and some of the former Soviet state nations such as Ukraine. Nevertheless, the effect of the actions of Russia goes far beyond its immediate neighbours. For instance, currently Western Europe has good reason to be concerned. Although it does not depend much on Russia as compared to former Soviet nations, Moscow's readiness to mercilessly employ its petrol power has created much concern among European Union Member States. The Union is therefore currently trying to persuade Russia to sign an Energy Charter restricting its influence. Other nations including the United States need to be worried by the Russian energy wealth increase and the manner in which this earned power is being employed.

17.4 Conclusions

The world is experiencing discriminative natural resource depositions. Some countries have more natural resource depositions compared to others. This has for a long time created disparity in different nations' economic positions, relations and political stability. A number of scarce resources are considered to be very essential and necessary for the survival of all in the world. For instance, the need for gas and oil seems basic to all nations for their normal operations. This has resulted in the development of international trade whereby those with more resources share with the rest of the world based on set financial agreements. Just like all other nations, Europe is experiencing disparity in the distribution of natural energy resources. Despite this, the continent cannot be considered among those with enough depositions to support their needs or even the needs of others. Therefore, European nations are among many other nations in the world that depend on Asian or North African nations for their natural gas and oil supply. The highest part of the European nations' natural energy resources is imported from Russia.

Europe imports natural gas and oil from Russia through pipelines that pass through other nations. This means that this trade involves not just the two nations that are directly involved, Russia and Europe, but also the nations where these pipes pass through. Therefore, the natural energy trade is a complex business that involves geo-political and geo-structural complexity that when not handled well can result in severe conflicts among nations. This aspect has invoked the development of a number of policies to define the relation among all the involved nations. Normally, these policies are created to reduce conflicts and to facilitate the solving of conflicts. Besides this, policies are also made considering complex interlinkages among different resources. In this case, there is a relation between different resources insofar as that, to extract one, one may need another one and, thus, their uses are highly influenced by how much the utilisation of one resource negatively affects the existence of another resource. Although this interdependency needs to be highly considered in policymaking, most policymakers find it too complex, and in this regard they normally consider one or more other resources while making these policies.

Europe has for a long time depended on Russia for its natural gas and oil supplies. However, with the current change of climate, technology and economy, the EU has established a number of policies that influence its relation with Russia. For instance, the recent climate policy of checking greenhouse gas effects discourages the use of fossil oil that emits more carbon oxides and other gases that are responsible for greenhouse gas effects. This move has made the country focus more on renewable sources of energy, and reducing the use of oil on the continent. Besides this, the current geopolitical conflict between Ukraine and Russia is clearly demonstrating the possibility of supply insecurity, and thus Europe is currently considering diversifying its sources of natural energy to ensure enough supply when Russia cuts its gas supply to the continent.

Global communication enhances interaction between nations and the development of good relations that promote unity and togetherness. In addition, global communication normally promotes trade among nations, enhancing a symbiotic relationship that promotes the different economic and general growth of a country by being able to relate with more people from different nations. This good relationship can be used to eliminate geopolitical conflict by enhancing peaceful negotiations and development of policies that are supported by all parties. Global communication promotes learning about other countries including their problems, their needs and their way of perceiving things, and thus it would be easy for policymakers to come up with policies that would be highly appreciated by all involved members. When a country has more than one relationship with another country in the same region, it should be easy for the two countries to create business relationships without initiating any form of conflict.

Notes

1 F. M. Verrastro, S. Ladislaw, M. Frank and L. A. Hyland (2010). *The Geopolitics of Energy Emerging Trends, Changing Landscapes, Uncertain Times*. CSIS Energy and National Security Programme. Accessed January 2014 from https://csis-prod.s3.amazonaws.com/s3fs-public/legacy_files/files/publication/101026_Verrastro_Geopolitics_web.pdf
2 M. Ratner, P. Belkin, J. Nichol and S. Woehrel (2013). *Europe's Energy Security: Options and Challenges to Natural Gas Supply Diversity*. Congressional Research Service 1–29, August 20, 2013. Accessed April 2014 from https://fas.org/sgp/crs/row/R42405.pdf
3 Ibid.
4 P. Stigson (2011). *The Resource Nexus: Linkages between Resource Systems*. Berlin: Elsevier.
5 C. A. Scott, S. A. Pierce, M. J. Pasqualetti, A. L. Jones, B. E. Montz and J. H. Hoover (2011). Policy and institutional dimensions of the water–energy nexus. *Energy Policy*, 39, 6622–6630.

6 M. Bazilian, H. Rogner, M. Howells and K. K. Yumkella (2011). Considering the energy, water and food nexus: Towards an integrated modelling approach. *Energy Policy*, 39, 7896–7906.

7 Ibid.

8 D. P. Giurco, B. C. Mclellan, D. M. Franks and T. Prior (2014). Responsible mineral and energy futures: Views at the nexus. *Journal of Cleaner Production*, 84.

9 P. Kratochvil and L. Tichy (2013). EU and Russia discourse on energy relations. *Energy Policy*, 56, 391–406.

10 S. Boussena and C. Locatelli (2013). Energy institutional and organizational changes in EU and Russia: Revisiting gas relations. *Energy Policy*, 55, 180–189.

Appendix A

Constructing a metadatabase from open data sources

I prefer databases that generally meet the following criteria:

* are publicly available and accessible;
* include a relatively comprehensive set of end-use measures commonly included in energy programmes;
* have been in use over several years or more;
* include data and/or algorithms for both energy (kWh) and demand (kW) savings, supply resources, occupant behaviour, weather, etc., along with sufficient detail on other key parameters and specifications;
* are well documented.

However, in reality things stand differently. The best way is to start from the question: 'What kind of data do we need?' Of course, this depends on the research questions you try to answer, and your aims and objectives.

For example, a database can be used to find relevant information for EU countries regarding demand, supply and trade with a structure such as: categories (social, economic, energy, environment); time (resolution; annual, quarterly, hourly, etc.); and space (by country, grid square, etc.).

Figure A.1 shows my initial attempt at a metadatabase constructed for EU countries.

The target was to create a database for the EU countries with the following in mind (main units TWh):

1 population
2 GDP
3 primary energy consumption
4 primary energy supply
5 oil production
6 oil imports
7 oil exports
8 oil consumption
9 oil supply
10 oil reserves
11 gas production
12 gas imports
13 gas exports
14 gas supply
15 gas reserves
16 gas consumption
17 coal reserves
18 coal production
19 coal consumption
20 electricity production
21 electricity imports
22 electricity exports
23 electricity consumption
24 electricity supply
25 share of renewables in electricity
26 solar consumption
27 wind consumption
28 geothermal, biomass and other consumption
29 CO_2 emissions.

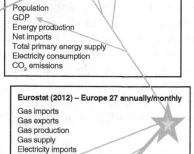

BP database (world)

Oil: proved reserves
Oil: proved reserves – barrels (from 1980)
Oil: production – barrels (from 1965)
Oil: imports and exports
Oil: production – tonnes (from 1965)
Oil: consumption – barrels (from 1965)
Oil: consumption – tonnes (from 1965)
Oil: regional consumption – by product
   group (from 1965)
Oil: spot crude prices
Oil: crude prices since 1861
Oil: refinery capacities (from 1965)
Oil: refinery throughputs (from 1980)
Oil: regional refining margins (from 1992)
Oil: trade movements (from 1980)
Oil: inter-area movements
Gas: proved reserves
Gas: proved reserves – bcm (from 1980)
Gas: production – bcm (from 1970)
Gas: production – bcf (from 1970)
Gas: production – Mtoe (from 1970)
Gas: consumption – bcm (from 1965)
Gas: consumption – bcf (from 1965)
Gas: consumption – Mtoe (from 1965)
Gas: trade movements pipeline
Gas: trade movements LNG
Gas: trade 2011–2012
Gas: prices
Coal: reserves
Coal: prices
Coal: production – tonnes (from 1981)
Coal: production – Mtoe (from 1981)
Coal: consumption – Mtoe (from 1965)
Nuclear energy: consumption TWh
   (from 1965)
Nuclear energy: consumption Mtoe
   (from 1965)
Hydroelectricity: consumption TWh
   (from 1965)
Hydroelectricity: consumption Mtoe
   (from 1965)
Renewables: other renewables
   consumption TWh (from 1990)
Renewables: other renewables
   consumption Mtoe (from 1990)
Renewables: solar consumption – TWh
   (from 1990)
Renewables: solar consumption – Mtoe
   (from 1990)
Renewables: wind consumption – TWh
   (from 1990)
Renewables: wind consumption – Mtoe
   (from 1990)
Renewables: geothermal, biomass and
   other – TWh (from 1990)
Renewables: geothermal, biomass and
   other – Mtoe (from 1990)
Renewables: biofuels production – Kboe/d
   (from 1990)
Renewables: biofuels production – Ktoe
   (from 1990)
Primary energy: consumption Mtoe
   (from 1965)
Primary energy: consumption by fuel
   type Mtoe
Electricity generation: TWh (from 1985)
Carbon dioxide emissions
Renewable energy: geothermal
   (installed capacity)
Renewable energy: solar (installed capacity)
Renewable energy: wind (installed capacity)

Enerdata database (world)

Total primary production
Total balance of trade
Total primary consumption
Energy intensity of GDP
Crude oil, NGL production
Crude oil, NGL balance of trade
Crude oil, NGL input to refinery
Oil products production
Oil products balance of trade
Oil products domestic
   consumption
Natural gas production
Natural gas balance of trade
Natural gas domestic consumption
Coal and lignite production
Balance of trade of coal and lignite
Coal and lignite domestic
   consumption
Electricity production
Electricity balance of trade
Electricity domestic consumption
Share of renewables in electricity
   production
Share of renewables in primary
   consumption
CO_2 emissions from fuel
   combustion
CO_2 intensity

IEA database (1) up to 2010

Population
GDP
Energy production
Net imports
Total primary energy supply
Electricity consumption
CO_2 emissions

IEA database (2) up to 2010

Energy production (Mtoe)
Net imports (Mtoe)
Total primary energy supply (Mtoe)
Net oil imports (Mtoe)
Oil supply (Mtoe)
Electricity consumption (TWh)
GDP (billion 2000 USD using exchange rates)
GDP (billion 2000 USD using PPPs)
Population (millions)
Industrial production index (2005=100)
Total self-sufficiency
Coal and peat self-sufficiency
Oil self-sufficiency
Gas self-sufficiency
TPES/GDP (toe per thousand 2000 USD)
TPES/GDP (toe per thousand 2000 USD PPP)
TPES/population (toe per capita)
Net oil imports/GDP (toe per thousand 2000 USD)
Oil supply/GDP (toe per thousand 2000 USD)
Oil supply/population (toe per capita)
Electricity consumption/GDP (kWh per 2000 USD)
Electricity consumption/population (kWh per capita)
Index of industry consumption/industrial production
   (2005 = 100)
Index of industry oil consumption/industrial
   production (2005 = 100)

Eurostat (2012) – Europe 27 annually/monthly

Gas imports
Gas exports
Gas production
Gas supply
Electricity imports
Electricity exports
Electricity supply

Figure A1 A metadatabase for EU countries.

Appendix B
A compilation of open data sources

There are lots of websites, reports, etc. that provide access to databases for free or under licence. I have compiled a list of various open sources at different levels. Please note the list is only a very tiny fraction of what is available online. The scope here is to show some examples of great initiatives worldwide.

General

- Openmod initiative's Wiki is a unique initiative worldwide. It is an Open Source and Open Data platform in energy modelling, where the user can create new articles and can register to edit and create pages. See http://wiki.openmod-initiative.org/wiki/Open_Models;
- Github archive – this is a meta list of public datasets on the hub: www.githubarchive.org. It includes data on climate/weather, networks, economics, energy, GIS, etc.;
- The Open Climability Suite is an open-source project. The software can be freely used under the terms of the GPL licence. See http://climability.org/demo/about.

Global level

- Renewables.ninja – generate wind and solar profiles from MERRA weather data globally;
- US DOE maintains a database of energy storage installations worldwide: www.energystorageexchange.org;
- Enipedia – a wiki-based collection of global power plant data run by Chris Davis (TU Delft);
- SciGRID – an open transmission grid topology developed by NextEnergy and derived from OpenStreetMap;
- IAEE EDS – a list of energy data links compiled by the International Association for Energy Economics;
- Data on Global Transmission Network – an open data source of the global transmission network based on OpenStreetMap: http://globalatlas.irena.org/News DetailPublic.aspx?id=2278 published by IRENA;
- IEA Energy Balances of Non-OECD Countries, including over 100 non-OECD countries and 11 regions;
- IEA World Energy statistics and balances;
- UN Energy statistics;

- World Bank Development Indicators contain statistical data for over 600 development indicators for over 200 countries and 18 country groups running from 1960 onwards, including data on social, economic, financial, natural resources and environmental indicators, etc.;
- Eurostat Harmonized Indices of Consumer Prices (HICP);
- HWWI Index of World Market Prices/Commodity Price Index.

National level

- UK Data Service (www.ukdataservice.ac.uk/get-data/key-data/international-macro-databanks), the largest collection of UK and international social, economic and population data;
- EnergyMap.info – a user-friendly database of all renewable-based power generators in Germany;
- DUKES – a list of all the power generation plants in the UK broken down by: operating company, station name, fuel, installed capacity (MW), start date and what category of English region or UK country it falls into;
- in line with ONS recommendations regarding presentation of subnational National Statistics, the following dataset, for 2009 data only, reflects the local government reorganisation operative from 1 April 2009: sub-national electricity consumption statistics: 2009 [filetype:xlsx filesize: 198.58Kb];
- electrical energy consumption profiles collected by the Florida Solar Energy Center (FSEC);
- the New Zealand Household Energy End-use Project (HEEP);
- the Canadian Residential Energy End-use Data and Analysis Centre (CREEDAC) Study and part of the UK Department of Trade and Industry (DTI) Photovoltaic Study.

Appendix C

A compilation of energy systems models available online

The following is a list with examples of models available online. Again, the list is not exhaustive.

- EnergyPLAN (www.energyplan.eu) is free to use, but there isn't a free way to examine and modify the code;
- OSeMOSYS is a free and open source software. See www.osemosys.org/getting-started.html;
- SciGRID Transmission Network Model (www.scigrid.de);
- Python for Power System Analysis (PyPSA) by the Renewable Energy Systems Group at FIAS. See https://github.com/FRESNA/PyPSA.

Index